This fourth volume of Colin Ronan's abridgement of Joseph Needham's monumental work is concerned with the immense advances made in early and medieval China in mechanical engineering. It covers the main sections of volume IV part 2 of the original treatise.

It discusses in simple but eminently readable terms the status of engineers, their tools and materials, then basic mechanical principles, followed by machinery powered by animals, man and even by steam, vehicles for land transport, six centuries of hidden clockwork, windmills and aeronautics. Since China was far ahead of the West in ancient and medieval times, this volume helps make clear the immense debt owed by Western civilisation to China. Such debt included the important mechanical principles of transforming rotary motion to a to-and-fro motion of a crank and vice versa. The Chinese invented the first efficient harness for horses, and the first mechanical clocks the world had ever seen.

THE SHORTER
SCIENCE AND CIVILISATION IN CHINA

COLIN A. RONAN
Lately of the Needham Research Institute

The Shorter
Science and Civilisation
in China

AN ABRIDGEMENT OF

JOSEPH NEEDHAM'S ORIGINAL TEXT

Volume 4

THE MAIN SECTIONS OF VOLUME IV, PART 2
OF THE MAJOR SERIES

ENGINEERS: THEIR STATUS, TOOLS
  & MATERIALS
BASIC MECHANICAL PRINCIPLES &
  TYPES OF MACHINES
LAND TRANSPORT
CLOCKWORK
WINDMILLS & AERONAUTICS

CAMBRIDGE
UNIVERSITY PRESS

CAMBRIDGE UNIVERSITY PRESS
Cambridge, New York, Melbourne, Madrid, Cape Town,
Singapore, São Paulo, Delhi, Tokyo, Mexico City

Cambridge University Press
The Edinburgh Building, Cambridge CB2 8RU, UK

Published in the United States of America by
Cambridge University Press, New York

www.cambridge.org
Information on this title: www.cambridge.org/9780521338738

First published 1994

A catalogue record for this publication is available from the British Library

Library of Congress Cataloguing in Publication data

ISBN 978-0-521-32995-8 Hardback
ISBN 978-0-521-33873-8 Paperback

CONTENTS

ILLUSTRATIONS

TABLES

PREFACE

In this, the fourth volume of the abridgement of Dr Joseph Needham's *Science and Civilisation in China*, we look at the astounding development of mechanical engineering in ancient and medieval China. We cover most of volume IV, part 2 of Dr Needham's original work. However, it is now desirable to keep each volume of this abridgement of approximately the same size; in consequence, and with Dr Needham's agreement, the description of hydraulic machinery has been held over until the next volume. There it will appear together with that part of hydraulic engineering concerned with control, construction and maintenance of waterways.

As in previous volumes in the abridgement, I am indebted to Joseph Needham for his encouragement and help; his advice has been invaluable. As in the previous volumes, this is no new edition. However, in the text the Pinyin romanisation of Chinese is given, but its use has now become so widespread that it is given first, with the modified Wade–Giles system in square brackets where desirable. The modified system has been continued in order to make it easy for readers to refer to Joseph Needham's original *Science and Civilisation in China*, should the need arise. Pinyin alone is given in the captions to and lists of illustrations.

My warmest thanks are due to Dr Simon Mitton and to Ms Fiona Thomson of Cambridge University Press for their ever ready help, and to Ms Helen Spillett not only for her careful checking of the Pinyin transliterations, but also to her and to Ms Sheila Champney for their admirable work as copyeditors, and to Ms Liz Granger for her excellent index.

The preparation of this volume has only been possible due to a grant generously provided by the Chiang Ching-kuo Foundation for International Scholarly Exchange in Taiwan. For this I am most grateful.

Hastings, East Sussex   Colin A. Ronan
September 1992

1

Engineers: their status, tools and materials

Mechanical engineering in China reached a very high stage of development at a time when Western engineering was still in a comparatively primitive state. Yet advanced though it became, it may be described as 'eotechnic', depending upon readily available naturally occurring materials. In its Chinese manifestation this meant primary reliance on wood, bamboo, stone and water. That is not to say that metals were unknown; indeed, they too were of great importance. From the Zhou [Chou] period (from the first millennium to the third century BC) when bronze was used for weapons, and in a refined form for gear wheels and crossbow-triggers in the Han (202 BC to AD 220), a time when cast-iron ploughshares were also in general use, and even steel-making was first practised, the Chinese used metals for what seemed to them appropriate purposes. In some respects, such as the mastery of iron-casting, and the first use and knowledge of zinc, the Chinese were much ahead of the Europeans. Yet it is true to say that most engineering constructions of any large size continued to be mainly of wood and stone. In this, change did not come until Western Renaissance technology spread over the Asian continent.

Yet if the Renaissance brought changes to East Asian technology, its practitioners were quite unaware of the origins of their subject. Hardly anyone in the Middle Ages would have noticed that technology had a history. Not until a literary quarrel in the sixteenth and seventeenth centuries AD, did it gradually dawn on historians that the ancient Romans did not write on paper, knew nothing of printed books, and used no collar harness, spectacles, explosive weapons or magnetic compasses. The disquiet caused by this realisation was partly the occasion of the controversy between the 'Ancients' and the 'Moderns', which took its place as an important aspect of the inevitable clash between the humanistic polymaths and the experimental philosophers. Among the protagonists of the Moderns, Jerome Cardan in 1550 signalised the compass, printing and gunpowder as the three inventions to which 'the whole of antiquity has nothing equal to show'. Almost a century later, in 1620, Francis Bacon in his *Novum Organum* was even more emphatic:

1

It is well to observe the force and virtue and consequence of
discoveries. These are to be seen nowhere more conspicuously than
in those three which were unknown to the ancients, and of which
the origin, though recent, is obscure and inglorious; namely,
printing, gunpowder and the magnet. For these three have changed
the whole face and state of things throughout the world, the first
in literature, the second in warfare, the third in navigation; whence
have followed innumerable changes; insomuch that no empire, no
sect, no star, seems to have exerted greater power and influence in
human affairs than these mechanical discoveries.

Bacon was mistaken in thinking that these inventions were inglorious, but
few writers at that time, and few historians later, recognised clearly their
non-European origin, or drew from the fact its full implications. As far as
mechanical engineering is concerned, something of all this will become evident in what follows.

THE NAME AND CONCEPT OF ENGINEER

A few words may not be out of place here regarding the origins of the terms
used for engineers in Western languages and in Chinese. To our minds the
word 'engine' has come to have so vivid and precise a meaning that it is hard
at first to remember that it derives from that quality of cleverness or ingenuity
which is (or was thought to be) inborn in certain people – 'ingenium', indwell-
ing genius, innerly generated. Since the derivatives of these roots were already
in common Roman use for expressing qualities of wit, craft and skill, it is
not surprising that 'ingeniarius', as a term in the more restricted sense, is
found in Europe with increasing frequency from the twelfth century AD
onwards. Not till the eighteenth was it freed from its primary military con-
notation. The course of events in China was not quite parallel to this.

From the earliest times the word *gong* [*kong*] (工) implied work as an
artisan, but technical as opposed to agricultural. This is perpetuated in the
modern term *gong cheng* [*kung chhêng*] (工程), the second of the two characters
having originally meant measurement, dimension, quantity, rule, examination, reckoning, etc. Other old words such as *ji* [*chi*] (機) (originally the
loom, the supreme machine) and *dian* [*tien*] (電) (originally lightning) came
in due course to do duty for mechanical and electrical devices respectively.
But none of their combinations, as applied to persons, is even medieval. The
really old term for artisan-engineer is *jiang* [*chiang*] (匠 or 匠), which is
probably a carpenter's square. One ancient oracle-bone form (𢦏) actually
shows a man holding a carpenter's, square *ju* (chü) (巨 or 矩). The word *gong*
itself also derives from a drawing of this instrument. It is safe to conclude

that in Chinese culture, primarily eotechnic as it was, engineering work *par excellence* was woodwork. The *Zhou Li* [*Chou li*] (Record of the Institutions of (the) Zhou [Chou] (Dynasty)) of Former Han times calls master-craftsmen *Guo Gong* [*Kuo Kung*] (國工) or Master Carpenters.

However, this is not the whole story. Technical ability which was particularly skilful and admirable was called *qiao* [*chhaio*] (巧). The radical character on the right here is interesting because it is related to a number of other forms of which the general meaning is 'that of breathing out'. Very familiar is the terminal expiatory word *xi* [*hsi*] (兮) used so much in Zhou and early Han poetry. Some relatives such as *hao* (号 or 號), to call out, are still in common use. The significance of the meaning of the word for engineering genius in Chinese, therefore, would be identical with that of Latin, but expressed in the opposite way, not emphasising that genius which was born *in*, but that which was manifested and breathed *out*.

Sometimes artisans and engineers were simply called 'makers' or 'doers', the term 'Makers or Doers' being applied, for example, as early as Qin [Chhin] times (third century BC) to officials in charge of artisans and workshops. The associations of Chinese terms for engineers and artisans seem, therefore, always to have been more civilian and less military than those used in the West.

ARTISANS AND ENGINEERS IN FEUDAL BUREAUCRATIC SOCIETY

In volume 2 of this abridgement the point was made (pp. 76 ff) that astronomers were part of the civil service. To some extent, and on a lower plane, artisans and engineers also participated in this bureaucratic character, partly because in nearly all dynasties there were elaborate Imperial Workshops and Arsenals. An additional reason was that, during certain periods at least, those trades which possessed the most advanced techniques were 'nationalised', as in the Salt and Iron Authorities under the Former Han (that is from 119 BC onwards). Also, as we shall see, there was a tendency for technicians to gather round the figure of one or another prominent official who encouraged them as his personal followers.

We are, of course, not here dealing with philosophers, princes, astronomers or mathematicians, the educated part of the Chinese population, but with those concerned with the obscurer expanses of the trades and husbandries. Certainly they applied scientific principles, whether or not always fully formulated, but now a new factor enters. We can no longer leave out of account the mass of the workers and the conditions under which they laboured. They were the human material without which the planners of irrigation works or bridges, or vehicle workshops, or even the designers of astronomical apparatus, could have done nothing, and not seldom it was from them that ingenious inventors and capable engineers rose up to leave particular names in history.

It is generally allowed that the most important document for the study of ancient Chinese technology is the *Kao Gong Ji* [*Khao Kung Chi*] (Artificers' Record) chapter of the *Zhou Li* [*Chou Li*]. Though the book in general is a Han compilation, it embodies a great deal of earlier date, probably from an official compilation of the State of Qi [Chhi] in the Warring States period (480 to 221 BC). Indeed, careful study indicates that much of it dates from not later than the second century BC and refers to the early part of the third century BC, as well as to still earlier periods. We are therefore dealing with a time when the Great Wall of Qin Shi Huang Di [Chhin Shih Huang Ti] was under construction, and at a period in the West when Euclid was busy at the new Library and Museum at Alexandria, and one of the ancient 'Seven Wonders of the World' – the giant lighthouse on the island of Pharos in the harbour there – was being built.

Its opening paragraphs are so interesting that they must be quoted extensively:

> The State has six classes of workers, and the hundred artisans form one of them.
>
> There are those who sit to deliberate on the Dao [Tao] (of Society) and there are others who take action to carry it on. Some examine the curvature, the form and the quality (of natural objects) in order to prepare the five raw materials (presumably metal, jade, leather, wood and earth), and to distribute them for making instruments (useful for) the people. Others transport things rare and strange from the four corners (of the world) to make objects of value. Others again devote their strength to augment the products of the earth, or to (weave tissues from) silk and hemp.
>
> Now it is the princes and lords who sit to deliberate upon the Dao, while carrying it into execution is the function of ministers and officials. Examining the raw materials and making the useful instruments is the work of the hundred artisans. Transportation is the affair of merchants and travellers, tilling the soil belongs to the farmers, and weaving is the office of women workers . . .
>
> Tools and machines were invented by men of wit (*zhi zhe* [*chih chê*]), and their traditions maintained by men of skill (*qiao zhe* [*chhaio chê*]); those who continue them generation by generation are called artisans (*gong*, [*kung*]). So all that is done by the hundred artisans was originally the work of sages. Metal melted to make swords, clay hardened to make vessels, chariots for going on land

and boats for crossing water – all these arts were the work of sages.

Now heaven has its seasons and earth has its *qi* [*chhi*] (local influences), particular stuffs have their virtues (*mei*) and particular workers have their skills; if these four things are brought together, something good comes out of it . . .

There now follow various examples, among them descriptions of excellent products from various regions, such as the knives of (the state of) Zheng (Chêng) and the double-edged swords from Wu and Yue [Yüeh]. Then, after remarks about the seasons, the text continues:

> Generally speaking, wood-working comprises seven operations, metal-working six, treatment of skins and furs five, painting five and polishing five, modelling in clay two. Woodwork includes the making of wheels, chariot bodies, bows, pikestaffs, house-building, cart-making, and cabinet-making with valuable woods. Metal-work includes forging (*zhu* [*chu*]), smelting (*ye* [*yeh*]), bell-founding, making measures, containers, agricultural implements and swords. Work with skins includes drying, making hide armour, drums, leather and furs. Painting includes embroidery in one or more colours, the dyeing of feathers, basketry and silk cleaning. Polishing includes the working of jade, the cutting and testing of arrows, sculpture and the making of stone-chimes. Modelling in clay includes the art of the potter and that of the tile-moulder.
>
> This last was the art most esteemed by the dynasty of Shun. The Xia [Hsia] gave first place to the art of house-construction, and that of the Yin (Shang) preferred the art of making vessels. But that of the Zhou [Chou] set highest the work of chariot body builders.

The whole passage shows something of the characteristic Chinese love of arbitrary systematisation, but is clearly based on fact. Moreover, in the last paragraph, mention of the favourite techniques of several dynasties is an example of their correlative thinking – the connections they formulated between things formed of similar elements in the basic Five Element chain. (See this abridgement, volume 1, pp. 153 ff)

Again, the four conditions of industrial production – season, local factors, virtues of materials and skill – given in the fourth paragraph provide interesting statements about animal and plant ecology, as well as providing a background to the Chinese care in siting industries according to the presence locally of coal or metallic ores, forests, water and so on. Admittedly, the descriptions of type of artificer seem meagre, far less complete than the tables of officials with their ranks and classes of assistants in other ministries. But this is because those for the Ministry of Works were in a section of the text now lost. What we have are mentioned in Table 47 (below), and we do know that the Imperial Workshops certainly produced all ceremonial objects, commodities of daily life, as well as vehicles and machines, required for the courts of the emperor and the princes. What is more, there could be no sharp distinction between such work and the manufacture of arms and equipment for the imperial forces. And when the salt and iron industries were 'nationalised', all the artisans concerned in them must also have come under immediate government control.

Nevertheless, it is safe to assume that when any large or unusually complex piece of machinery was constructed (e.g. the early water-mills) this was done either in the Imperial Workshops or under the close supervision of important provincial officials. For all dynasties seem to have had Imperial Workshops, so in general one can conclude that a considerable proportion of the most advanced technologists in all ages in China were either directly employed by, or under the close supervision of, administrative authorities forming part of the central bureaucratic government.

Yet not all were employed in this way. The great majority of artisans and craftsmen must always have been connected with small-scale family workshop

Table 47. *Trades and industries described in the Kao Gong Ji chapter of the Zhou Li*

(A) WORKERS IN STONE AND JADE

Jade workers[a]	*yu ren*	玉人
Stone carvers	*diao ren*	雕人
Stone chime makers	*qing shi*	磬氏

(B) CERAMICS WORKERS

Potters	*tao ren*	陶人
Moulders (tiles)	*fang ren*	㼧人

(C) WOOD WORKERS

Arrow makers	*jie ren*	桺人
	shi ren	矢人
Bow makers[b]	*gong ren*	弓人
Cabinet makers in valuable woods[c]	*xi ren*	梓人
Weapon handle makers	*lu ren*	盧人
Surveyors, builders and carpenters	*jiang ren*	匠人
Agricultural implement handle makers, *see* Cartwrights		

(D) CANAL AND IRRIGATION DITCH BUILDERS (and hydraulic engineers in general)

Hydraulic workers[d]	*jiang ren*	匠人

Table 47. *contd*

(E) METAL WORKERS (*gong jin zhi gong*)		
'Lower alloy' founders[e]	*zhu shi*	築氏
'Higher alloy' founders[f]	*ye shi*	冶氏
Bell-founders	*fu shi*	鳧氏
Measure makers	*li shi*	㮚氏
Plough makers	*duan[g] shi*	段氏
Sword-smiths	*tao shi*	桃氏
(F) VEHICLE MAKERS[h]		
Wheelwrights	*lun ren*	輪人
Master wheelwrights	*guo gong*	國工
Body makers	*yu ren*	輿人
Shaft and axle makers	*zhou ren*	輈人
	zhu ren	軸人
Cartwrights[i]	*che ren*	車人
(G) ARMOURERS (of hide, not metal)		
Cuirass makers	*han shi*	函氏
(H) TANNERS		
Tanners	*wei ren*	韋人
Skinners	*bao ren*	鮑人
Furriers	*qui ren*	韗軍人
(I) DRUM-MAKERS	*yun ren*	韗人
(J) TEXTILE, DYEING, AND EMBROIDERY WORKERS		
(*hua i chih shih* 畫繢之事)		
Feather-dyers	*zhong shi*	慌氏
Basket-makers	*kuang ren*	筐人
Silk-cleaners[j]	*mang shi*	鍾氏

[a]Much information about the forms of the various ceremonial pieces, but hardly a word about the techniques of working.

[b]A long and elaborate section, which makes no mention of the crossbow.

[c]Mainly musical instruments and cups. They had a foreman or manager, *zi shi*.

[d]This section contains valuable information on irrigation canals, cf. *Science and Civilisation in China* volume IV part 3.

[e]'Lower alloy' bronze was a 3 parts copper: 2 parts tin mixture, said here to be used for writing knives.

[f]'Higher alloy' bronze was a 3 parts copper: 1 part tin mixture, said here to be used for arrow-heads, lance-heads, etc.

[h]Their measurements and dimensions are related to standard weapon lengths.

[i]Also make handles for agricultural implements.

[j]Those who remove the gum from the natural silk.

production and commerce. Indeed, the largest part was played by handicraft production independently undertaken by and for the ordinary people. As a result, particular localities derived fame from skills which tended to concentrate there, such as the lacquer-makers of Fuzhou [Fuchow], the potters of Jingdezhen [Ching-tê-chen], or the well-drillers Ziliujing [Tzu-liu-ching] in Sichuan [Szechuan]. Over and over again, people of the countries bordering China expressed their respect for the artisans of China. Nor did they hesitate to ask for them when circumstances permitted. Thus in 1126 AD, when nomadic tribes from Manchuria – the Jurchen (Jin [Chin]) Tartars – besieged the Song [Sung] capital at Kaifeng [Khaifêng], they demanded from the city all sorts of craftsmen, including goldsmiths and silversmiths, blacksmiths, weavers, tailors and even Daoist [Taoist] priests. Again, when the famous Daoist Qiu Changchun [Chhiu Chhang-Chhun] made his famous journey from Shandong [Shantung] to Samarkand at the request of Genghis Khan in 1221, he met Chinese workmen everywhere. In Outer Mongolia they came in a body to meet him, with banners and bouquets of flowers; when he got to Samarkand he found numbers more. And as late as 1675, we find a Russian diplomatic mission to Beijing [Peking] officially requesting that Chinese bridge-builders be sent to Russia.

Where did the Chinese inventors and engineers come from? They were commoners (*xiao min* [*hsiao min*]) (小民), and for the ancient philosophers 'menial men' (*xiao ren* [*hsiao jen*]) (小人) as opposed to 'magnanimous quasi-aristocratic scholarly official men' [*jun zi* [chün tzu]) (君子). Having surnames they were of the *bai xing* (*pai hsing*) (百姓) (the 'old hundred families'), and belonged to the *bian min* (*pien min*) (編民) (registered people). But whatever the extent of government-organised production from time to time, the State relied upon an inexhaustible supply of unpaid labour in the form of the *corvée* (*yao,* 傜 and 繇), (*yi* [*i*], or *yu* [*yü*], 役); (*gong yu* [*kung yü*], 公役). In Han times every male commoner between the ages of twenty (or twenty-three) and fifty-six was liable for one month of labour service a year, unless belonging to some specially exempted group. Technical workers certainly performed these obligations in the Imperial Workshops or in the factories of such enterprises as the Salt and Iron Authorities. These organisations were never staffed by slaves, and as time went on there naturally grew up the practice of paying dues in lieu of personal service so that a large body of artisans were 'permanently on the job'. In Yuan times (AD 1271–1368) government artisans were distinguished from military artisans, though both received pay and rations, differing from the private civilian artisans, the services of whom, however, could be requisitioned from time to time. Artisans were always spared by the Mongol (Yuan) conquerors, who assembled them in government factories. As in the Song [Sung], artisans could not be conscripted for any service other than their trade. During the Ming (AD 1368–1644), a rota of technical

Fig. 247. A late Qing [Chhing] representation of artisans at work in the Imperial Workshops. From the *Qinding Shu Jing Tu Shuo [Chhin-Ting Shu Ching Thu Shuo]* (The *Historical Classic* with Illustrations). Qing (edition by imperial order 1905).

corvée labour in government factories according to regular registers became prominent.

Artisans who were regular commoners were by no means the lowest social level in ancient and medieval China. Below them came a number of 'depressed classes'. The general term for these was *jian min* [*chien min*] (賤民) 'base' or 'ignoble' people in contradistinction to the *liang ren* [*liang jen*] (良人), the 'commoners'. The ignoble were not slaves, though slavery did exist in ancient and medieval China; it was primarily domestic in character and originally essentially a punishment. As far as unfree artisans were concerned, their proportion was less than 10 per cent of the whole mass of workers in the crafts and techniques.

In the Han, like the Zhou [Chou] before it, male slaves were those condemned to penal servitude, female ones to grain-pounding. Though war captives sometimes suffered this fate, in ancient times the main source was certainly from criminals and their whole families. A great many were owned directly by the imperial State, but the line is hard to draw because State slaves were often given to high officials and nobles in reward for services or as presents. Their status, which they shared with no other groups, was that of property; but disabilities such as fixation of domicile, restrictions of marriage outside their own group, and restraint from change of occupation were also suffered by the 'ignoble' class. Indeed, the general idea of permanence of technical trade was a long-lasting one, but convicts whether men or women were 'enslaved' for a term of years, or for life, with no descendants. As in the ancient West they were sent to the mines.

We might well expect technicians among the slaves, the descendants of convicts, since special training and long experience would be more natural among people spending a lifetime in servitude, and indeed it seems likely that the life of a government slave tradesman was often considerably more comfortable, and certainly more secure, than that of a 'free' yeoman artisan.

As to the skills possessed by slaves, a remarkable document from 59 BC, purporting to be a purchase contract, has survived since the Han. Apart from a slave's labour in garden and orchard, he was supposed to plait straw sandals, hew out cart shafts, make various pieces of furniture and wooden clogs, whittle bamboo writing tablets, twist rope and weave mats – but besides all this he was to make knives and bows for sale in the neighbouring market. Thus it appears that among the slaves there were always those who were skilled craftsmen, though never as many as among the commoners who had to act as *corvée* labour.

The Chinese also used the term *tong* [*thung*] (and 童), which may perhaps best be translated as "serving-lads". As this term had a strong ndustrial undertone, one is almost tempted to view them as bonded apprentices, perhaps for a long term of 'educational' years. Indeed, in the third and second centuries

BC, we know of such *tong* being employed in iron foundries, as well as in centres concerned with extensive domestic production of, for example, fine textiles. Gradually *tong* became associated with the terms *bu qu* [*pu chhü*] (部曲) and *ke ren* [*kho jen*] (客人) or *ke nü* [*kho nü*] (客女), 'bearers' and 'guests'. These were 'retainers' who could be transferred from one master to another, but could not be bought or sold. Artisans were always to be found among them.

Yet whatever characteristics medieval Chinese labour conditions may have possessed, they proved no bar to a long series of 'labour-saving' inventions altogether prior to those arising in Europe and Islam. Lugging and hauling were avoided whenever possible. Moreover, in all Chinese history there is no parallel for the slave-manned oared war-galley of the Mediterranean – land-locked though most of the Chinese waters were; sail was the characteristic motive power throughout the ages. And when the water-mill appeared early in the first century for operating the bellows used in metallurgical work, the records distinctly say that it was considered important as being not only cheaper than man-power or animal-power, but also more humane. Evidently shortage of labour is not in every culture the sole stimulus for labour-saving invention. Indeed in Chinese culture inventions were never rejected because of fear of technological unemployment. Yet in Europe it seems to have been a perennial fear, and industrialisation had to go forward in the teeth of it.

With this background, we can now turn our attention briefly to a survey of the social groups from which inventors and engineers appear to have originated. Let us use five divisions: first, high officials, the scholars who had successful and fruitful careers; second, the commoners; third, members of the semi-servile groups; fourth, those who were enslaved; and fifth, the rather significant group of minor officials, namely those scholars unable to make their way upwards in the ranks of the bureaucracy.

First, then, the high officials. Here we may take Zhang Heng [Chang Hêng] who flourished in AD 120, and Guo Shoujing [Kuo Shou-Ching] (flourished AD 1280) as outstanding representatives of the type, both of whom we have already met in volume 2 of this abridgement. Zhang Heng was the inventor of the first seismoscope in any civilisation, and the first to apply motive power to the rotation of astronomical instruments; besides this he was a brilliant mathematician and designer of armillary spheres. Guo Shoujing, more than a millennium later, was an equally good mathematician and astronomer but also a most distinguished civil engineer who constructed the Tonghui [Thung Hui] Canal and planned most of the Yuan Grand Canal. Both of these men occupied the position of Astronomer-Royal, but in addition Zhang Heng became President of the Imperial Chancellery, while Guo Shoujing was Intendant of Waterways and Academician. Two entirely comparable figures were Su Song [Su Sung] and Shen Gua [Shen Kua], whom we shall mention in a moment. All four had the good fortune to find appreciation for their scientific and technical talents in their age – others were not so lucky.

Sometimes ability of this kind accompanied military gifts and offices. The Qin [Chhin] general Meng Tian [Mêng Thien], who flourished 221 BC, comes to mind. He accomplished the fusion and extension of previously existing walls to form the Great Wall, and built a 950 kilometre (590 mile) road from Ningxia [Ninghsia] to Shanxi [Shansi]. Du Yu [Tu Yü] (flourished AD 270) was prominently associated with the spread of hydraulic trip-hammers and establishing multiple geared water-mills for cereal grinding, as well as throwing pontoon bridges across great streams such as the Yellow River.

Sometimes, again, provincial officials are credited with important technical developments. Thus the introduction of the water-powered metallurgical blowing engine is attributed to Du Shi [Tu Shih] who was prefect of Nanjing [Nanking] in AD 31, and its further spread, so vital for the technical mastery of cast-iron production at this early period, was due to another governor, Han Ji [Han Chi], who was prefect of Loling. The most obvious case is that of Cai Lun [Tshai Lun], who began as a confidential secretary to the emperor, was made Director of the Imperial Workshops in AD 97, and announced the invention of paper in 105.

Chinese princes and the remoter relatives of imperial houses were favoured with leisure because, though generally well educated, they were in most dynasties ineligible for the civil service yet disposed of considerable wealth. Though the great majority did little for posterity, a memorable few devoted time and riches to scientific pursuits. Here we shall mention only one or two, because for various reasons they tended to interest themselves in astronomical, biological or medical rather than in technical or engineering matters. But the Han noble Liu Chong [Liu Chhung], prince of Huainan [Huai-Nan] (flourished AD 173) was the inventor of grid sights for cross-bows, and a famous shot with them as well. In the Tang [Thang] we meet with Li Gao [Li Kao], prince of Cao [Tshao] (flourished 784), interested in acoustics and physics, but prominent here because of his successful use about this time of treadmill-operated paddle-boat warships.

Curiously, it seems quite exceptional to find an important engineer who attained high office in the Ministry of Works, age-old though this department was in the Chinese bureaucratic pattern. Perhaps this was because the real work was always done by illiterate or semi-illiterate artisans or master-craftsmen, who could never rise across that sharp gap which separated them from the 'white-collar' literati in the offices of the Ministry above. Perhaps they sometimes felt that they could get on with the job much better if the administration upstairs were poets and courtiers not too uncomfortably familiar with the tools and materials of the trade and its mystery.

However that may be, there were exceptions. Yuwen Kai [Yüwên Khai] (flourished AD 600), chief engineer of the Sui dynasty for thirty years, carried out irrigation and conservation works, superintending the construction of the Tong Ji Qu [Thung Chi Chhü], part of the Grand Canal, built a large sailing-carriage, and with Geng Xun [Kêng Hsün] (to whom we shall return shortly) devised the standard steelyard clepsydra (volume 2, p. 155 of this abridgement) used throughout the Tang [Thang] and Song [Sung]. He also built new capital cities at Chang'an [Chhang-an] in 583 and Loyang [Loyang] in 606, and made a wooden model of the cosmological Ming Tang [Ming Thang] temple. For many years Minister of Works, his earlier post had been Director of the Architectural and Engineering Department of the Imperial Palaces, so Yuwen Kai must have been a real expert in all the mechanical and constructional arts of his time.

During the Ming dynasty the way upwards for artisans to enter the administrative grades of the Ministry of Works seems to have been more open. For instance, several woodworkers and joiners, notably Kuai Xiang [Khuai Hsiang] (flourished 1390 to 1460), Cai Xin [Tshai Hsin] (flourished 1420) and Xu Gao [Hsü Kao] (flourished 1522 to 1566), who all showed merit as builders and architects, succeeded in this way, and the last named rose to be President of the Ministry.

Extending this list would bring in many more names of men of high official rank. This would partly be due to the enormous social importance of hydraulic engineering works in Chinese society, a fact which rendered this skill – which will be described in volume 5 of this abridgement – always highly honourable among scholars and administrators tending to purely literary accomplishments. But it would also be due to a tendency readily discernible for technicians to cluster in the entourage of a distinguished civil official, who acted as their patron. Here we find the examples of Su Song [Su Sung] and Shen Gua [Shen Kua] instructive. Su Song (flourished 1090), a most distinguished official who served as Ambassador and President of the Ministry of Personnel, was responsible as we shall see (p. 225), for the construction of the great astronomical clock tower at Kaifeng and for the greatest treatise on time measurement of the Chinese Middle Ages. To accomplish this, he surrounded himself with a remarkable band of engineers and astronomers, whose names he preserved and transmitted. Shen Gua (flourished 1080), equally gifted and equally successful, was Ambassador and Assistant Minister of Imperial Hospitality, but we think of him chiefly as the author of the most interesting and many-sided scientific book of the Song period. It is in this work that we find the best authentic statement of the beginnings of printing in China, and so we are introduced to that great inventor Bi Sheng [Pi Shêng], a 'man in hempen cloth' (i.e. a commoner, or one not dressed in silk) who first devised, about 1045, the art of printing with movable type. 'When Bi

Sheng died', says Shen Gua, 'his fount of type passed into the possession of my followers, among whom it has been kept as a precious possession until now.' Thus we have a striking glimpse of the entourage of technicians which an enlightened official could gather round him. Finally, well-known officials would be likely to figure largely in any survey because there were always other reasons for the insertion of their biographies in the dynastic histories.

With Bi Sheng we come now to the commoners, the *liang jen*. Men whose names alone we know can probably be safely placed in this group. Ding Huan [Ting Huan] (flourished AD 180) renowned for his pioneer use of the gimbals suspension (p. 145) and for his construction of rotary fans and ingenious lamps, is termed simply a 'clever artisan'. Again, Yu Hao [Yü Hao] (flourished AD 970), that brilliant designer and builder of pagodas, was but a Master-Carpenter, and his celebrated *Mu Jing [Mu Ching]* (Timberwork Manual) was assuredly dictated to a scribe. Sometimes we have only the surname of a valued man, e.g. Lacquer-Artisan Wang from Suzhou [Suchow], who about AD 1345 devised dismountable boats and collapsible armillary spheres for the imperial court, and some fifteen years later even rose to be Intendant of one of the Imperial Workshops. Sometimes we do not even have the surname – an omission which makes one wonder whether such men were members of the servile or semi-servile groups in which surnames were not customary.

In the category of commoners, we should also place minor military officers, and certainly Daoist and Buddhist monks. Among the former was Tang Dao [Thang Tao], the gallant defender of the city of Hanyang [Han-yang] in Hubei [Hu-pei] province during the years AD 1127 to 1132 when it was repeatedly attacked by the Jurchen Jin [Jurchen Chin]. Together with the civil magistrate Chen Gui [Chhen Kuei], apparently an equally inventive mind, he used successfully for the first time a new device called the 'fire-lance' – a reversed gunpowder rocket held in the hands and employed as a defensive shock weapon. Though the barrels were not of metal, and the composition spattered material rather than propelling it, this was undoubtedly the real origin of all barrel guns and cannon.

In view of the close association between Daoism and the technical arts in ancient China, one would expect to find more Daoist inventors in the Middle Ages than have appeared. Nevertheless it is not difficult to name some. Tan Qiao [Than Chhaio] experimented with lenses about AD 940 (see volume 2 of this abridgement, pp. 360 and 361), while five centuries earlier, in AD 450, Li Lan was responsible for a long line of development of 'stop-watch' clepsydras (water-clocks) using vessels of jade and mercury as the flowing material. He also developed larger clepsydras which weighed the water remaining in them by a steelyard balance. Among the Daoists we must not forget a woman, Jia Gu Shan ('The Valley-Loving Mountain Immortal'), who engineered mountain roads in Fujian province about AD 1315.

On the whole Buddhists were more illustrious as technicians in the Middle Ages. One outstanding example is Yi Xing [I-Hsing], the greatest astronomer, mathematician, and instrument-designer of his age, the eighth century AD, whom we have already met in volume 2 of this abridgement. Though debarred from all official rank, he was a member of the College of All Sages and the most trusted scientific man at court for nearly twenty years. Of greater engineering interest perhaps are several monks, notably Fa Chao [Fa-Chhao] (flourished 1050) and Dao Xun [Tao-Hsün] (flourished 1260), who built many of the wonderful megalithic stone bridges across the rivers and estuaries of Fujian province. Since the length of the beams is frequently only just below that at which the stone breaks under its own weight, it would seem that strength-of-materials tests were carried out at this time.

And now we come to the exceptional cases, men who came down in history as brilliant technologists yet whose social standing in their own time was very low indeed. The only one of clearly semi-servile rank in our registers is Xindu Fang [Hsintu Fang] (flourished AD 525). In his youth he entered the household of a prince of Northern Wei, who had long been collecting many pieces of scientific apparatus – from armillary spheres to clepsydras and windgauges – and who also possessed a very large library. As a man of known scientific skill, Xindu Fang's position must have been like that of expert artisan and technical advisor to his patron, the prince. It seems that the prince intended to write certain scientific books with the help of Xindu, but owing to political and military events felt obliged to flee to the south in 528, so that Xindu had to write the books himself. After this he remained in seclusion, probably in poverty, till he was called to another court, that of the 'king-maker', Gao Zu [Kao Tsu]. Here he was a 'retainer' and served as Estate Agent, a post which may have exercised his talents in surveying and architecture. A man of humble and abstracted disposition, he was nevertheless consulted by fairly high officials, designed strange rotary fans in connection with the tuning of pitch pipes and even engaged himself in calendar reform.

Examples of technologists who were positively slaves seem also rare, but there is Geng Xun [Kêng Hsün] (flourished AD 593). He began working under a governor of Lingnan [Ling-Nan], but when his patron died, he joined some tribal people in the south and eventually led them in an uprising. When this was defeated and Geng captured, the general Wang Shiji [Wang Shih-Chi], realising his technical ability, saved him from death and admitted him among his slaves. Here his position was not so low that he could not receive instruction from an old friend Gao Zhibao [Kao Chih-Pao], who had become Astronomer-Royal, and as a result of this Geng built an armillary sphere or celestial globe rotated continuously by water-power. It is interesting to find that the emperor rewarded him for this achievement by making him a government slave and attaching him to the Bureau of Astronomy and Calendar.

We now reach the last of our groups of technicians, and one of the most numerous, namely that of the minor officials – men who were sufficiently well educated (even if of lowly origin) to enter the ranks of the bureaucracy, but whose particular talents or personalities frustrated all hopes of a brilliant career. These were the kind of men who could have become famous in science or engineering in a post-Renaissance world. Take Li Jie [Li Chieh], for instance (flourished AD 1110), the man who building on the earlier works of others produced the greatest definitive treatise of any age on the millennial tradition of Chinese architecture. This volume, *Ying Zao Fa Shi* [*Ying Tsao Fa Shih*] (Treatise on Architectural Models) was written when he was only an Assistant in the Directorate of Buildings and Construction, though he did in the end attain the Directorship. For Li Jie we have complete biographical details, but in a hundred others there is no such record. An elaborate specification for hodometers or distance-measuring vehicles prepared by Lu Daolong [Lu Tao-Lung] in AD 1027 has come down to us, but no information whatever about the life of this engineer.

Sometimes such Chinese engineers found they could make a better career in the service of foreign dynasties, where the pressure of conventional literary culture was less and ingenuity could find spontaneous if unsophisticated admiration and support. Thus around AD 340, Xie Fei [Hsieh Fei] and Wei Mengbian [Wei Mêng-Pien] both served Shi Hu [Shih Hu], the king of the Hunnish Later Zhao [Chao] dynasty. Wei was his Director of Workshops, and together with Xie produced south-pointing carriages (see p. 189), wagon-mills, revolving seats and floats with complicated mechanical puppets, as well as other items, for all of which Shi Hu had a particular liking. A century later, another 'nomadic' dynasty, the Yan, which was a prototype of the Mongol dynasty which spanned the thirteenth and fourteenth centuries, employed Zhang Gang [Chang Kang], the most famous military engineer of his time (*c.* 410), a great expert on crossbows and perhaps the first inventor of the type using multiple-springs which were so characteristic of Chinese artillery in the days before gunpowder was used in war during the eleventh century. Zhang was famous also for his knowledge of fortifications and the ways of attacking them. In the end, however, he returned to Chinese allegiance and joined the founder of the Liu Song [Liu Sung] dynasty.

Some ingenious men held posts and followed occupations which seem to have been quite unsuited to their talents. Yan Su [Yen Su] (flourished AD 1030) was a Leonardo-like figure – scholar, painter, technologist and engineer under the Song emperor Ren Zong [Jen Tsung]. He devised a type of water-clock with an overflow tank which remained standard for long afterwards, invented special locks and keys and left specifications for devices like hodometers and south-pointing carriages. His writings included treatises on time-keeping and on the tides. Yet most of his life was spent in provincial administrative posts and though he did become an Academician-in-Waiting of the Long-Tu [Lung-Thu] Pavilion, he never rose above Chief Executive Officer of the Ministry of Rites and had no connection with the Ministry of Works or other technical directorates. Again, Han Gonglian [Han Kung-Lien] (flourished AD 1090), principal collaborator to Su Song [Su Sung] in applied mathematics during the construction of the greatest clock-tower in Chinese history, was an Acting Secretary in the Ministry of Personnel when Su Sung found him and as far as we know, that is where he permanently remained.

To end this section, the best thing we can do is to present a translation of part of an essay written in the third century AD by the philosopher and poet Fu Xuan [Fu Hsüan] on his friend the engineer Ma Jun [Ma Chün] (flourished 260), perhaps one of the most interesting documents on the social history of ancient and medieval Chinese technology found by Dr Needham.

> Mr Ma Jun . . . came from Fufeng [Fu-fêng] and was a man of wide renown for his technical skill. In his youth he travelled into Henan [Honan] but (at that time) he did not yet realise his own talent. Even then his powers of exposition fell far behind his mechanical ingenuity, and I doubt if he could express half of what he knew. Although he had a literary degree he remained poor. He therefore thought of improving the silk loom, and thus at last without need of explanations the world recognised his outstanding skill . . . The old looms had fifty heddles (harnesses to guide threads) and fifty treadles. Some even had sixty of each. Mr Ma . . . changed the design in such a way that it had only twelve treadles. Thus strange new patterns in many wonderful combinations were made, by the inspired conception of the inventor, all arising easily and naturally; . . . But alas, how could he hope to explain to people the (principles of the) improvements which he had made?
>
> Mr Ma, being a Policy Review Adviser one day fell into a dispute at court with the Permanent Counsellor Gaotang Long [Kaothang Lung] and the Cavalry General Qin Lang [Chhin Lang] about the south-pointing carriage. They maintained there had never been any such thing and that the records of it were nonsense. Mr Ma said: 'Of old there was. You have not thought the matter out. It is really not far from the truth.' But they laughed . . . To this Mr Ma replied: 'Empty arguments with words cannot (in any way) compare with a test which will show practical results. All this was reported by Gaotang and Qin to the emperor Ming Di [Ming Ti],

whereupon Ma Jun received an order to construct such a vehicle.
And he duly made a south-pointing carriage. This, was the first of
his extraordinary accomplishments. But again it was almost
impossible to describe (the principle of it) in words. However,
henceforth the world bowed to his technical skill.

Fu Xuan next goes on to describe further accomplishments. He then
continues:

> (At this time) there was at court a notable scholar, Master Pei (Pei
> Zi [Phei Tzu]). When he heard (of Ma's inventions) he laughed and
> mocked him with difficult questions. Mr Ma stammered and could
> not give (satisfactory) replies. As Master Pei could not get the
> essential ideas from Ma's explanations, he continued to discredit
> him. I myself said to Master Pei on one occasion: 'Your great
> merit is of course your eloquence, but where you fall short is
> technical skill. Now this is Mr Ma's strong point, but he is not a
> good talker. For you to attack his inability to express himself is
> really not fair. On the other hand when you argue with him about
> those technicalities in which he excels, there must be points which
> we cannot expect to understand. His special talent is a very rare
> one in the world. If you insist on raising difficulties about these
> matters which are so hard to expound, you will go far astray from
> the truth. For Mr Ma's gifts are all of the mind and not of the
> tongue. He will never be able to reply (to all you ask of him).'
>   Later on I gave an account of my talk with Master Pei to the
> Marquis of Anxiang [An-Hsiang]. (Unfortunately) the Marquis
> agreed with Master Pei. So I said: 'The sages selected talent in
> accordance with ability, and did not carelessly entrust people with
> the management of affairs. Some obtained distinction because of
> their spirit, others because of their eloquence. The former had no
> need of words, for their quality was revealed by their sincerity and
> virtuous deeds, while the latter had to argue about right and wrong
> in order to demonstrate their greatness ... There were also those
> (who obtained distinction) because of their political ability, like Ran
> You [Jan Yu] and Jilu [Chi-Lu], and those who showed literary
> brilliance, like Ziyou [Tzu-Yu] and Zixia [Tzu Hsia). Thus even
> the all-understanding sages used trials and tests in the selection of
> important personnel. Ran You was tested with politics, and Ziyou
> and Zixia with learning. Why not apply such tests to lesser
> persons? What is the point of saying that the principles of things
> cannot be exhausted in words, and that there can be no end to

discussions about the universe, when the truth can so easily be verified by experiment? Mr Ma is proposing to construct ingenious equipment for the country and the army. All you need to do is to give him ten measures of wood and a couple of workmen, and you will soon know who is right. Why should it be so difficult to (get permission) for (an official) test when the experiment is readily made? To discredit extraordinary ability in other people with light words is like (someone) who would impose his own wisdom on the affairs of the universe instead of handling the endless difficulties in accordance with the Dao [Tao]. This is the path to destruction. Must you refuse to make use of Mr Ma because not everything that he has said proved to be right? How do you expect less well-known technical men to come forward? Among the mass of the people jealousy among competitors and mischief among colleagues is unavoidable. Hence (wise) rulers pay no attention to such things and base their judgment on tests . . .

After this the Marquis of Anxiang saw my point, and talked over the matter with the Marquis of Wu'an. Nevertheless, the (official) tests were never ordered. (Alas, when the government) neglects to arrange such simple trials for a man like Mr Ma whose skill was well known, what hope is there for lesser gems to be brought to light? (I hope) that the rulers in days to come will use this case as a mirror . . . When (authorities) employ personnel with no regard to special talent, and having heard of genius neglect even to test it – is this not hateful and disastrous?

After this remarkably scientific and experimental cry from the third century AD, further comment would be superfluous, and we leave the last word to Fu Xuan.

TRADITIONS OF THE ARTISANATE

Only one thing more remains before we can get to the bench, the foundry and the field. Any picture of an eotechnic artisanate would be incomplete without some allusion to its own traditions. About these something has already been said in this abridgement (volume 1, p. 105), while the names of primary inventors mentioned in the *Yi Jing* [I Ching] (Book of Changes) in its Great Appendix compiled about the second century BC were also given (p. 185). These sages, introducers of the plough, the cart, the boat, the gate and so on, were doubtless those to whom Confucius was referring in his famous but truncated aphorism: 'The great inventors were seven in number.' Here we need only add to the above the specifically Daoist [Taoist] traditions enshrined in the *Lie Xian Zhuan* [*Lieh Hsien Chuan*] (Lives of the Famous Hsien (Immortal Beings)). A book of this title certainly existed in the Later Han (first century AD), but traditionally it was attributed to Liu Xiang [Liu Hsiang] (77 BC to 6 BC) who is supposed to have based it on a set of pictures. The text as we now have it was fixed only as late as 1019 AD, when books about the Fathers of Daoism were first printed, but internal evidence shows that some portions of it go back to dates such as 35 BC and 167 AD. As we should expect from the nature of ancient Daoism, many of the immortals in this book have close connections with the mechanical trades. Chi Song Zi [Chhih Sung Tzu] and Ning Fen Zi [Ning Fen Tzu] are concerned with the mastery of fire involved in metallurgy and ceramics, cast iron has a patron in Tao Angong [Thao An-Kung], and mirror polishing with mercury in Mr Fuju [Fu-Chü]. The Daoist patron of makers of mechanical toys was certainly Ge You [Ko Yu], who animated one and rode away on it, while Lu Pi Gong [Lu Phi Kung] was the great magician of bridges, ladders and galleries. Even the reel or the small windlass of the fishing rod had its spirit in Dou Ziming [Tou Tzu-Ming].

The greatest of all the tutelary deities of artisans was Gongshu Pan [Kung-shu Phan] (Fig. 248). In spite of the fact that much of what was handed down about him is clearly legend, there is no reason to doubt his real existence in the state of Lu (hence his other name Lu Ban [Lu Pan]) in the fifth century BC, and we shall meet him here from time to time in connection with kites and other devices. He lives in proverbs, for instance 'brandishing one's adze at the door of Lu Ban', which is as much as to say, in our less elegant idiom, 'teaching one's grandmother to suck eggs'. And as so often in Chinese culture, where everything tends to be one of a pair, like parallel hanging scrolls, Gongshu Pan has a companion Wang Er [Wang Erh], the semi-legendary inventor of curved chisels and graving tools, who may well have been a master of wood-carving contemporary with him.

Here is the place to mention a curious little work called the *Lu Ban Jing* [*Lu Pan Ching*] (Lu Ban's Manual), which circulated widely in the recent past among China's craftsmen, and some description of it is appropriate. Its author and compiler and its editors are quite dateless, but much of its content is so archaic that one gains the impression of dealing with material some of which might well go back at least to the Song [Sung]. Anything so traditional will always be hard to date.

The book opens with a series of illustrations (e.g. Fig. 249). These show joinery operations, sawyers at work, and various kinds of houses, bridges and pavilions. Then there follows, after a legendary biography of Gongshu Pan, a mass of detail about the cutting of timber in forests, the making of granaries, bell-towers, summer-houses, furniture, wheelbarrows, the square-pallet chain pump, the piston bellows, the abacus and many other things. Precise specifica-

Fig. 248. Paper block-print icon of Gongshu Pan. In black on yellow paper, with decorative bands of colour in pink, green, mauve and red. Such icons were commonly pasted on the walls of workshops with incense-sticks burning before them. Attendants in the foreground bear the tools of the trade, those behind technical treatises. In accordance with the bureaucratic character of Chinese culture, Gongshu Pan, like all other tutelary deities, is enthroned as a magistrate or governor. Above him the inscription says: 'Master Gongshu of Lu, our Teacher from of Old'.

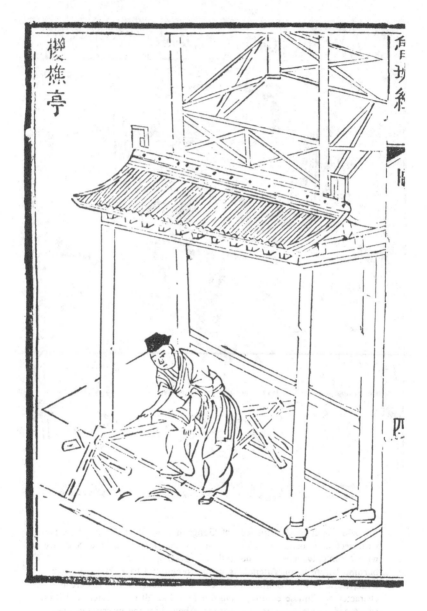

Fig. 249. An illustration from the *Lu Ban Jing*. Using a drawknife in the timberwork of a watch-tower under construction.

tions are interspersed with samples of charms and appropriate sacrifices, and lore about lucky and unlucky days; this last was more than just superstition for it had a social function in a culture which had not borrowed from Israel the institution of a weekly day of rest for toilers. As the book proceeds, the magical element increasingly predominates over the technical; there is a 'physiognomy' of buildings, and much on exorcism, luck-bringing incantations and other protective magic. The whole work, therefore, is a unique piece of traditional technology and folklore, and though not recognisable in any of the booklists in the dynastic histories, one can guess it has a pretty long past.

The most outstanding characteristic of Lu Ban's children down to modern times was the fact that they worked by knack, by rule of thumb, by an inherited slowly evolved tradition. Such became evident in our discussion of the artisanate of the ancient Chinese world. The Daoist philosophers, particularly interested in techniques not transmissible by words, were always giving examples of this incommunicable yet learnable skill – swordsmiths, arrowmakers and wheelwrights, for instance. Two immortal stories remain of them): Ding [Ting] the butcher, who cleaved his bullock carcasses according to the structure and pattern of the Dao, and Bian [Pien] the wheelwright who told Duke Huan of Qi that he would be better employed in learning the trade of governing from the people than in poring over Confucian treatises.

The justified veneration of 'knack' and personal flair arose largely in those long ages when materials and processes were not really fully under control. In a time such as our own, when a whole factory can operate almost automatically, it is hard even to imagine the lot of the technician who had to carry out procedures of which there was no scientific understanding. Deviations from the normal which would spoil the results desired could only be detected by the man whose observational faculties were extremely alert. Courses of action once found to succeed could only be repeated by the man whose memory was good and whose dexterity could rise to the occasion time after time. 'Materials as variable as wood and potter's clay and the crude uncomprehended metals could only be worked by the man who learnt from decades of experience to know the signs, the 'smell', the physiognomy, of the materials suitable for his purpose. There was many a failure and disappointment, but some psychological help was forthcoming from myth and legend, and many of the Daoist metallurgists engaged in rites of purification and self-discipline before beginning their operations. The craftsman could not express his procedures in logical terms. In fact he could not explain at all; he could only show.

Thus the transmission of the crafts from one generation to another naturally involved a total education of the body and spirit of the learner. Apprenticeship was subjective and personal, not a matter of the intellectual understanding, not at all the appreciation of mathematical functions describing the behaviour of deeply analysed physico-chemical entities. Yet to some extent the skill of the artisans was handed down orally in the widespread and invariable practice of 'learning by rote' mnemonic rhymes. This is marked not only in engineering and building literature, which is relatively small, but also among the alchemists and physicians, whose books are extremely numerous. For the tricky practical procedures of the medieval chemist nothing was more natural. Even some scholars, despairing of being able to produce accurate maps of the heavens, recommended the learning of the stars by heart. Of course in China, all this was part of a culture in which the normal method of approaching the study of the classics was learning by heart in school before any explanations were given.

But it must be remembered that in ancient and medieval times (and indeed until the last century) most of the craftsmen remained illiterate. There was therefore not only the basic intellectual difficulty of coining new technical terms, but also serious obstacles to writing them down when they were adopted. Shipbuilding is among the most complex of the mechanical arts, yet one of the deepest modern students of this subject in China found in thirty years' experience that hardly any of the best shipwrights whom he met could write. What is more, there are many technical sea terms for which no Chinese characters exist at all. Technical books did exist, but Dr Needham has found only one treatise with adequate diagrams, and that was never printed, nor is it very old. Indeed, it is worth remembering that in modern engineering parlance 'a Chinese copy' still means a copy of a machine or of some component part made by eye, measurement, or tradition, without any diagrams or drawings.

In one sense this was summed up as late as 1942 by G. Sturt who, writing on the work of a wheelwright commented in his book *The Wheelwright's Shop*:

> With the idea that I was going to learn everything from the beginning I put myself eagerly to boys' jobs, not at all dreaming that, at over twenty, the nerves and muscles are no longer able to put on the cell-growths, and so acquire the habits of perceiving and doing which should have begun at fifteen. Could not Intellect achieve it? In fact, Intellect made but a fumbling imitation of real knowledge, yet hardly deigned to recognise how clumsy in fact it was. Beginning so late in life I know now that I could never have earned my keep as a skilled workman.

Later, referring to the good sense handed down through generations of wagon-builders in Surrey, he commented that their knowledge:

> . . . was set out in no book. It was not scientific. I never met a man who professed any other than an empirical acquaintance with waggon-builder's lore.

And a little later, after describing things to be done, Sturt writes:

> This sort of thing I knew, and in vast detail in course of time; but I seldom knew why. And that is how most other men knew. The lore was a tangled network of country prejudices, whose reasons were known in some respects here, in others there, and so on . . . The whole body of knowledge was a mystery, a piece of folk knowledge, residing in the folk collectively, but never wholly in one individual.

Such is the background of medieval craftsmanship, in East as well as West. Out of it grew all those great inventions and innovations which preceded the breakthrough of the Renaissance, when the technique of invention and discovery was itself discovered. It is the scene on which the drama now opening was played.

TOOLS AND MATERIALS

In entering upon the present subject it is first necessary to say something briefly about the tool-chest of the Chinese mechanic (see Fig. 247, p. 9). The task is rendered difficult by the fact that no one, either Chinese or European, has brought together a critical discussion of Asian tools of all ages, taking into account both the ethnology of the subject and confronting archaeological objects with relevant texts. It has not even been done for our own civilisation. Here all that can be offered are a few notes and references which may help to orientate our minds.

Certainly, with the exception of shearing and possibly punching, Neolithic man had discovered all the basic mechanical principles embodied today in powerful machines for changing the volume and form of matter. The difference between the far past and the present lies mainly in the application of sources of power far exceeding that of human muscle, and controlled to an ever increasing extent by various technical refinements. The few remarks which have to be made may conveniently be arranged according to the classification of mechanical operations given in Table 48.

It is curious that in spite of the great social significance of tools and machines, only ten out of the 214 root forms (radicals) in the Chinese language represent them, with an additional three radicals for weapons. Dr Needham suggests that possibly this reveals lack of technological interest on the part of Confucian codifiers and lexicographers.

The history of hammers has been considered by a few scholars, and much information about the various kinds of hammers, mallets, sledgehammers etc., traditionally in use among Chinese artisans, was collected by Rudolph Hommel over half a century ago. He found they made great use of the

Table 48. *Classification of mechanical operations for changing the volume and form of matter*

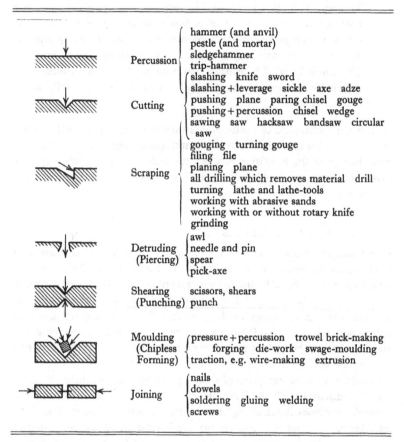

	Percussion	hammer (and anvil) pestle (and mortar) sledgehammer trip-hammer
	Cutting	slashing knife sword slashing + leverage sickle axe adze pushing plane paring chisel gouge pushing + percussion chisel wedge sawing saw hacksaw bandsaw circular saw
	Scraping	gouging turning gouge filing file planing plane all drilling which removes material drill turning lathe and lathe-tools working with abrasive sands working with or without rotary knife grinding
	Detruding (Piercing)	awl needle and pin spear pick-axe
	Shearing (Punching)	scissors, shears punch
	Moulding (Chipless Forming)	pressure + percussion trowel brick-making forging die-work swage-moulding traction, e.g. wire-making extrusion
	Joining	nails dowels soldering gluing welding screws

principle of flexible handles for heavy hammers, to add to the force of the impact and reduce the sting on the hands. When the handle of a hammer is pivoted at some point along its length, and the head raised and allowed to fall consecutively either by human power or water power, this tool has reached the level of a machine – the tilt-or trip-hammer (Fig. 250). It seems that this did not occur in Europe until the fifteenth century AD, yet it goes well back into the Han in China, as shown by numerous models of tilt-hammers operated by the weight of a man, and found in the tombs of the period. These may be compared with the oldest drawing which we have of such a device, which appears in the *Gena Zhi Tu* [*Kêng Chih Thu*] (Pictures of Tilling and Weaving) of 1210 AD (see Fig. 299, p. 113), where it is seen for removing the husks from rice.

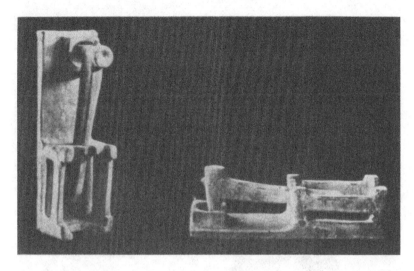

Fig. 250. Han model of man-powered tilt-hammer in iridescent green glazed pottery.

As might be expected, cutting edges in China had manifold uses. Besides their use by workers in wood and metal, there was a range of agricultural implements – the sickle, the sharp-pointed spade, the ploughshare and so on, together with military weapons such as the sword. The carpenter's plane was always used in China by pushing away from the worker's body, whereas Japanese and Korean planes are used in just the opposite way, pulled not pushed, perhaps because their users work on the floor while the Chinese work at a bench.

The combined operations of cutting, sawing and scraping go back to Neolithic times and were used in all civilisations. In China the saw achieved many subtle developments and a common form of it is the bow-saw or framed pit-saw for cutting trees into planks. In such saws the teeth are inclined starting from the middle of the blade in such a way that they point in opposite directions towards the ends; thus two sawyers, one above the wood and one below; do equal amounts of work. Sometimes logs are sawn horizontally, the sawyers walking slowly from end to end on each side (Fig. 251). Full stretching of bow-saws is effected by tightening the cords on the other side of the central pole of the frame with a toggle stick (Fig. 252). Such saws were used also in the West in Hellenistic times (about 300 BC to AD 300).

Since metals can be shaped and cut while in the hot plastic state, the saw plays a much smaller role in metal-working than in carpentry, and the file, which in a sense corresponds to the plane, a correspondingly greater one. The Chinese file differs somewhat from that commonly used in the West. It has

Fig. 251. Sawyers in the Chong qing shipyards in 1944. Part of a
photograph from Cecil Beaton's *Chinese Album*, London, 1945.

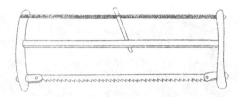

Fig. 252. Frame- or bow-saw with toggle stick. From *Tu Shu Ji Cheng* (Imperial Encyclopaedia), Qing, AD 1726.

a pointed section or tang at both ends, one to receive the wooden handle, and the other to carry a longer shaft of wood which fits through a ring-topped spike or eye-bolt (Fig. 253) within which it slides. This guide-rod gives a helpful down-leverage. While Western artisans hold the article to be filed firmly in a vice and work all round it with the file, the Chinese system is to have the file oscillating in a more or less fixed path, and the article is turned round under it as desired for filing.

All grindstones and whetstones belong in the same category as files. Grinding must have reached a high level of skill among the early Chinese because of jade-working among them (see volume 2 of this abridgement, pp. 317 ff). Hommel found no evidence of the use of rotary grindstones; all those that he saw were rectangular blocks worn to a concave upper surface. But complete absence of rotary ones seems difficult to believe, because the rotary disc-knife in jade-working goes back to the third century BC at least. Moreover, there is a mental connection between the disc-knife and the edge-runner mill (p. 124) as shown by the fact that the earliest term for the former which has come down to us is the same as for the latter, with the addition of 'abrasive' as an adjective.

Drilling must be considered a branch of scraping or grinding since material is actually removed from the hole made. This technique brings up the question of the origin of the driving-belt, and takes us therefore immediately into a fundamental engineering problem. Continuous rotary motion can only be attained by the aid of the crank (in one form or another) or the driving-belt, but for many purposes, of which drilling is one, a reciprocating rotary motion is sufficient. It has been said that the Chinese did not make use of continuous rotary motion, but this is a misstatement in view of the antiquity of the crank in China (p. 74) and machines such as the square-pallet chain-pump. If they had not been masters of it, the continuous and carefully adjusted turning of armillary spheres (pp. 230 ff) with the aid of water-power would not have been possible. Nevertheless, it does not seem that the artisans themselves applied a continuous drive to their drills, for even now the reciprocating drive remains in general use, as it is so widely in all eotechnic cultures and civilisations.

Fig. 253. Apprentices using the Chinese double-tanged file.

The simplest form of reciprocating rotary motion is that seen in the shaft-drill, the shaft of which is merely rotated back and forth between the palms of the hands. The ancestor of the driving-belt comes in when a strap or thong is wound round the shaft and each end pulled alternately (similar to the single motion of spinning top) by one man (Fig. 254); a second holds the drill steady. Here the belt is not continuous, but it becomes so in a sense, when its two ends are connected by a piece of wood, as in the case of a bow and its bowstring. If now the bowstring is wound once round the drill axis a powerful reciprocating motion will be imparted to the drill each time the bow is swept back and forth. Probably paleolithic, it is found among many eotechnic peoples still, and was a well-known tool in Ancient Egypt. It still exists in England in certain trades. A Chinese carpenter's bow-drill is shown in Fig. 255. Somewhere very long ago, no one knows where or when, the idea originated of making a hole in the centre of the bow and having it rise and fall alternately at right angles to the drill axis with a pumping motion, hence the term pump-drill. Fig. 256 shows a modern Chinese brass-smith's pump-drill.

Fig. 254. The strap-drill in use.

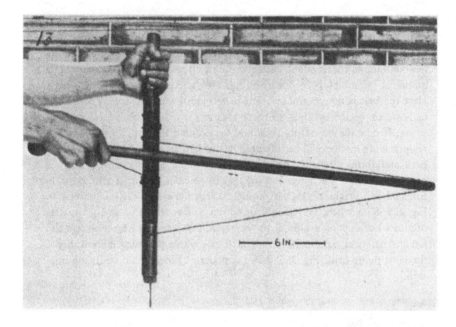

Fig. 255. The bow-drill in use.

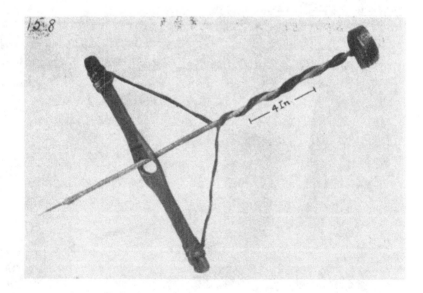

Fig. 256. The pump-drill, used by brass-smiths.

The most spectacular application of drilling in Chinese culture was the art of deep drilling, practised especially in Sichuan [Szechuan] province, where there are vast deposits of brine and natural gas. Techniques were developed by the time of the Early Han (third century BC), but details must be discussed elsewhere in this abridgement.

When a drill is mounted horizontally upon a frame, and is made to bear upon its working end not a boring tool but the article on which a chisel or knife is designed to work, then the lathe has come into existence. Treadles operated by the feet easily replace the two hands of the assistant operator, and this simple form of the machine is still common among Chinese artisans (Fig. 257). By a simple improvement, one end of the drive belt can be attached to a springy pole above the lathe, thus releasing the worker's feet and permitting more concentration on the work – such pole-lathes persisted in general use well into the eighteenth century in Europe, and are known in many cultures. The statement that this type of lathe never developed in China should be accepted with caution, since the use of springy bamboo laths was

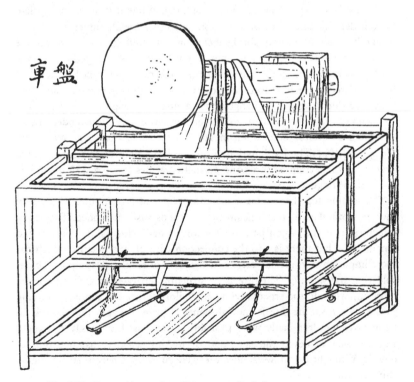

Fig. 257. Alternating-motion Chinese treadle lathe.

always very characteristic of Chinese technique as, for example, in one of the two basic types of looms. Little is known as to the antiquity of the lathe in China, but it is not likely to be later than Hellenistic times, when Egypt obtained it from the Greeks.

Shears, scissors and punches form a separate category since the material is attacked from both sides at once. Shears in which the tangs of the iron blades formed a continuous spring were common throughout the Roman Empire, but there seems no reason at present to suppose that the Han people got them from their Western contemporaries. A pair of such shears is to be seen in tomb frescoes of Song [Sung] times (fifth century AD) in Henan [Honan] province. A study of scissors from a Tang [Thang] dynasty tomb (seventh to tenth centuries AD) suggests that they derived from pivoting together two knives of a shape common in the Han. Of course, scissors have an obvious connection with instruments for holding small objects, such as pincers and forceps; and since the use of chopsticks goes back so far in Chinese civilisation – to the fourth century BC at least – there would seem more probability that if any travel took place, it was from east to west.

The last basic type of technique is moulding. While it is more appropriate to consider the casting of bronze, iron and other metals under metallurgy in another volume of this abridgement, an outstanding Chinese technique involving moulds is that of house-building with walls of stamped or rammed earth; this goes back to the Shang time (second millennium BC), which will be considered in the next volume. However, wire-drawing involves moulding because a ductile metal is changed by drawing it through a hole (Fig. 258). A suggestion that the Chinese could not draw iron wire until modern times is refuted by a clear description in 1637 in the *Tian Gong Kai Wu* [*Thien Kung Khai Wu*] (The Exploitation of the Works of Nature); this concerns the process in connection with making magnetic needles (see Fig. 173 in volume 3 of this abridgement).

The joining together of materials is a branch of a technique which may involve a number of the fundamental methods just mentioned. Projecting structures play the largest part in woodwork – nails, dowels, the mortice and tenon, etc. – but there is also the connection with glue; a practice which has more importance in metal-working in the form of similar techniques such as soldering and welding. The projecting structure separated from its base is the most important in the arts of sewing and tailoring, needles and pins deriving presumably from the thorns and fish-bones of primitive ages and peoples. In China wrought-iron nails were probably always used sparingly by wood-workers, who achieved the same end by making wood or bamboo pins or dowels. Wrought-iron dowels, however, were considered necessary in Chinese shipbuilding.

Last but not least in importance for the artificer were the tools of measure-

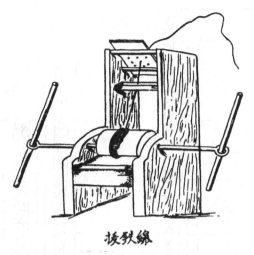

Fig. 258. Apparatus for wire-drawing.

ment. The oldest and simplest were the stretched string or plumb-line, the water-level, the measuring-rule, the compasses, the carpenter's square, and the balance or steelyard. Referred to already in volume 2 of this abridgement (pp. 330 ff), the square and compasses go back so far as to appear in the purest mythological material, for they form the traditional emblems of the organiser gods. Indeed, so familiar were the tools of measurement that even Confucian scholars drew upon them for metaphors. For example, the *Xun Zi* [*Hsün Tzu*] (The Book of Master Xun) from 240 BC says:

> When the plumb-line with its ink is truly laid out, one cannot be
> deceived as to whether a thing is straight or crooked. When the
> steelyard is truly suspended, one cannot be cheated in weight.
> When the square and compasses are truly applied, one cannot be
> mistaken as to squareness and roundness. So when once the
> great-souled man has investigated rightness of conduct, he cannot
> be deceived by what is false.

Examples of more complex measurements are frequent in illustrations and descriptions; for instance, testing the strength of cross-bows, or of specific gravities. The most striking measuring tool from old China, however, is perhaps the adjustable outside caliper gauge with slot and pin, which looks very like a modern adjustable spanner without the worm. A remarkable example dating from 9 AD, is shown in Fig. 259. No engineering tool of this kind

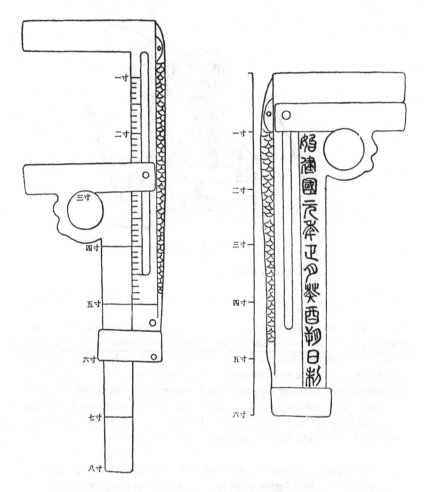

Fig. 259. Adjustable outside caliper gauge with slot and pin, graduated in inches and tenths of an inch (2.5 mm); in bronze. Self-dated at AD 9 by the inscription, which reads: 'Made on a *gui-you* day at new moon of the first month of the first year of the Shijianguo reign period'. This remarkable measuring tool is thus of the short-lived Xin [Hsin] dynasty, intermediate between the Earlier and Later Han.

appears to be known in Europe before the late fifteenth century, the time of Leonardo da Vinci, who sketched something similar.

One cannot quit the subject of tools and their work without alluding to one particular characteristic Chinese material, bamboo, which was not so readily available for other civilisations. A very large number of species are indigenous to China, probably the commonest being members of the genus *Phyllostachys*. Available literature on putting this wonderful material to use is very sparse, but by way of illustration of some of the techniques, Fig. 260 shows the detail of a litter chair. Such constructions make slight use of pegs or lashings, but one bamboo pole is made to hold another tight by chamfering and bending at a right angle, or even doubled back. Furniture, scaffolding and other erections make use at a more neotechnic level of exactly the same principles. Yet delicate works of art in bamboo are often described. The springiness of bamboo laths is also utilised to the full in all kinds of bow and bowstring

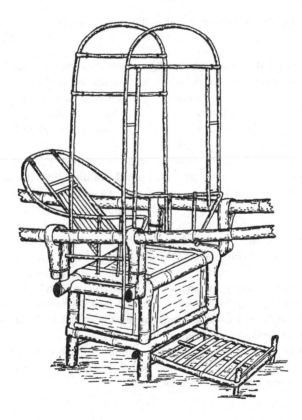

Fig. 260. This litter chair is an example of the clever traditional use of bamboo by Chinese artisans.

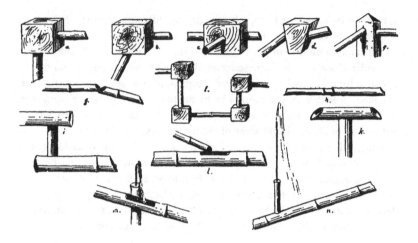

Fig. 261. Ways of using and jointing bamboo for pipes and conduits.

devices, but the remarkable tensile strength of bamboo was not fully realised until recent times, when experiments made by the Chinese Air Force Research Organisation showed that ply-bamboo of formidable qualities could be made by uniting layers of woven laths with glue. But all through the centuries this property had been used in practice to the full in bamboo cables and ropes for many purposes.

In another direction, bamboo with its internal partitions or 'septae' removed, forms a natural pipe and exerted a cardinal influence on East Asian invention. In earliest times it offered itself as a material for flutes and pipe-like musical instruments, which deeply moulded the development of Chinese acoustics throughout the ages. It was also used in piped water installations, and bamboo tubing was used by the Chinese in alchemy and the beginnings of chemical technology (Fig. 261). Cut along its length it served for light tiles on roofs and every sort of simple channel. It also fulfilled its most fateful destiny by becoming the ancestor of all barrel-guns. And lastly let it not be forgotten that bamboo-shoots were found to make excellent eating.

In what follows wood and bamboo are the most prominent materials for the construction of machines in ancient and medieval China, only certain essential parts being made of bronze or iron. The fact is that nowhere in the world was iron used for machinery before an industrial age began to demand long-enduring machines on a large scale.

2

Basic mechanical principles

We now cross the borderland from tools to machines. This vast subject will be dealt with in the following way: first comes a general discussion of the basic elements of which all machines are built up; second (pp 101 ff), we must consider the fundamental types of machines depicted in Chinese traditional illustrations, as well as provide a brief account of the coming of the Jesuits to China; third (pp. 155 ff), our understanding will be enlarged by the literary evidence, after which we shall be in a position to adopt firmer views about the probable times and places of origin of the various engineering techniques than those accepted provisionally at the outset. At the start of the second stage there will be some account of mechanical toys and automata, partly because they embodied nearly all of the basic mechanical principles, and partly because we have in China little or no pictorial evidence concerning them. At the opening of the third stage, there will be an account of vehicles, which can be differentiated rather sharply from all stationary engines. It will then prove convenient to arrange the third stage following a classification of power sources – animal, water descent and wind pressure – though hydraulic and civil engineering will be deferred until volume 5 of the abridgement. The fourth stage will deal with the early history of clockwork, and this volume will conclude with a few notes on vertical and horizontal mountings, on the windmill and, finally, the prehistory of aeronautical engineering.

Chinese engineering, with all its prodigious achievements, worked by a practical approach which overcame a general lack of theory of the subject, at least for a very long time. Moreover, we must not forget the factors of the social environment. In that Chinese society which could never of itself generate the equivalent of the Western Renaissance and the emergence of modern science, there was nevertheless something at work which encouraged mechanics and engineers to brilliant practical achievements, achievements which neither the men of the Hellenistic world, nor the inhabitants of medieval Europe before the time of Leonardo, attempted or attained. Such a contrast requires an explanation, but this is not yet the place for it.

A century and more ago Franz Reuleaux, a French engineer, wrote: 'A machine is a combination of resistant bodies so arranged that by their means the mechanical forces of Nature can be compelled to do work accompanied by certain determinate motions.' At about the same time, Robert Willis, an English physicist and archaeologist, amplified the first phrase: 'Every machine will be found to consist of a train of pieces connected together in various ways so that if one be made to move, they all receive a motion, the relation of which to that of the first is governed by the nature of the connection.'

One would not at first sight expect to find this thought in ancient China, but actually it is very old, and common to both the Chinese and the Greeks. Moreover, analogies made by early Greek philosophers between the world-machine and human machines were also echoed in China. And when the English physician William Harvey, in the sixteenth century AD compared the valves of the heart with the valves of a pump, he was in line with a view long held by the Chinese, for there was a Daoist conviction that the living body was a self-acting organic automaton, with sometimes one part in charge, sometimes another.

Heron of Alexandria (flourished AD 62) was apparently the first to classify machine elements; in his view they numbered five: the wheel and axle, the lever, the pulley, the wedge and the endless screw or worm. This was unsatisfactory, but explainable because it was in a book dealing only with the lifting of weights. Other early classifications were Indian. Here we shall deal with machine elements under the following heads: (*a*) levers, hinges and linkwork, (*b*) wheels, gear wheels, (*c*) pulleys, driving-belts and chain-drives, (*d*) crank and eccentric motion, (*e*) screws, worms and vanes of helical shape, (*f*) springs and spring mechanisms, (*g*) conduits, pipes and siphons, and (*h*) valves, bellows, pumps and fans.

LEVERS, HINGES AND LINKWORK

Of the lever and its great early application the balance, something has already been said (volume 2 of this abridgement, pp. 332 ff). There we saw that the Mohist engineers of the third century BC must have been acquainted with most if not all of the equilibrium principles stated by their Greek contemporary Archimedes. In the centuries immediately following, this understanding of the lever was put to good use in China in the making, almost on a mass-production scale, of crossbow triggers. These mechanisms, which involved intricate bent levers and catches, were beautiful and delicate bronze castings, and deserve the more extended description to be given in another volume of this abridgement concerned with military technology. On a scale much larger, and with the use of timber, the lever had also been employed from an earlier date in the swape or 'shaduf', a counterweighted bailer bucket.

In China the trip-hammer was an important application of the lever-press, and heavy loads tended to be hoisted by combinations of levers rather than by pulley tackle. But the most elaborate use of levers by the Chinese was undoubtedly in textile machinery, where levers and connecting rods were united with treadles to form complicated linkworks. Indeed, the Chinese were far ahead of the West in loom construction, perhaps symbolised by the fact that the Chinese word for loom, *ji* [*chi*], implies that it is the machine *par excellence*.

The assembly of such a linkwork, however, involves the use of hinges or movable joints. Essentially the hinge is a pin and two hooks, and both these structures were readily available in all antique civilisations. For doors or windows the pin tended to be long and the 'leaves' (the Chinese use the same term) broad and flat; for links in rods the pins would be short and the hook-tails elongated. Relevant in this connection are the curious bronze hooks on the ends of poles found in Loyang tombs of the sixth century BC; these seem to have been used for setting up easily dismountable booths or tents. There

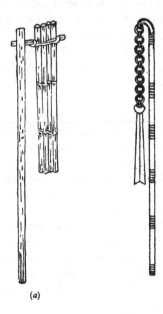

(a)

Fig. 262. Flails as examples of linkwork and chain connection. (*a*) Farmer's flail from the *Nong Shu* (Treatise on Agriculture of AD 1313. (*b*) Iron war-flail from the *Wu Jing Zong Yao* (Collection of the Most Important Military Techniques) of AD 1044. One of the names for the latter, the 'iron crane bird's knee', came to be used as a technical term in eleventh-century AD mechanical engineering for all kinds of combinations of rods and chains in linkwork.

were prominent uses of links in agriculture and war as well as in textile technology; first the link-flail and war-flail (Fig. 262), secondly the component parts of an efficient horse harness, which though not connected at their junctions by pins (being of leather), nevertheless performed the office of a linkwork system.

Perhaps more characteristic of European ideas of Chinese civilisation was the collapsible umbrella or parasol, working by means of sliding levers still familiar in everyday use. While sun-shades were common in Greek and Roman daily life, and certainly go back to Babylonian times, they were not generally collapsible. We have, however, an indication that the principle of the collapsible Chinese parasol was used in AD 21, for in that year Wang Mang had a very large one made for a ceremonial four-wheeled carriage. The mechanism is said to have been a secret one, but in the second century AD the commentator Fu Qian [Fu Chhien] adds that his umbrellas all had bendable joints enabling them to be extended or retracted. Collapsible umbrella stays of Wang Mang's time have been recovered from a tomb at Lo-Lang in Korea, but the system must go back much earlier, for similar objects of Zhou [Chou] date (sixth century BC) from Loyang have been discovered (Fig. 263).

Mastery of the art of collapsibility achieved some remarkable successes in the Yuan [Yüan] period (AD 1271–1638) which are referred to in a curious entry in the *Shan Ju Xin Hua* [*Shan Chü Hsin Hua*] (New Discourses from the Mountain Cabin) of 1360. Here Yang Yu [Yang Yü] tells us that:

> a lacquer worker of Suzhou [Suchow] named Wang, in the Zhi-Zheng [Chih-Chêng] reign-period (1341 onwards) made a boat of ox-hide, covered inside and out with lacquer. It was dismountable into parts, and was brought to Shangdu [Shangtu] (the Manchurian summer capital of the Yuan emperors), where it was rowed about on the Luan River. No one in Shangdu had ever seen such a thing before.
>
> Artisan Wang also constructed an armillary sphere by imperial command. This was collapsible, too, which was very convenient for storage. It was an instrument the skilful construction of which surpassed human imagination. One could really call this man talented. Now (1360) he is Intendant in charge of one of the (Imperial) Workshops.

However original the boats of Artisan Wang may have been, there had been successes in the twelfth century AD, and possibly going back to the Tang [Thang] (seventh to tenth centuries AD). In any case, the principle of collapsibility seems to have been a recurrent one.

Fig. 263. Bronze castings of complex design from the sixth century BC (Zhou period) excavated at Luoyang. (*a*) Bronze socketed hinge with locking slide-bolt; (*b*) bronze rebated socket couplers with holes for tenons and stanchion pins, some using the principle of the bayonet catch; (*c*) bronze rebated hinged corner-fitting, socketed to hold wooden members. Cast bronze tubular connecting joints for assembling the rods or canes of ancient Chinese carriage canopies; (*d*) two six-way branching holders of the fourth century BC.

WHEELS AND GEAR-WHEELS, PEDALS AND PADDLES

With the ultimate origin of the wheel we are not directly concerned, for by the time at which we can begin to speak of the history of technology in China, the Shang period (about 1500 BC), the chariot with its wheels had already been introduced. There seems to be some doubt about its origin; the traditional view was that the wheel first appeared in Sumeria about 3000 BC, but in the 1920s Sir John Marshall's excavations at Harappa in the Mohenjo-Daro civilisation (Indus Valley) found models of similar dating. All the same, the earliest examples of spoked wheels belong to northern Sumeria, and are datable about 2000 BC.

The most ancient Sumerian and Chaldean chariots were strange saddle-shaped structures for 'riding' borne on two wheels, the platform chariot not appearing until after 2500 BC. This was the form in which it spread to Egypt and Shang China. Possibly the use of rollers for moving heavy weights on quaysides and building-sites preceded the use of wheels, but it is a common misconception that they were used for moving heavy statues; for such work it seems that what were first used were 'sliders' on which a lubricant was poured and thus these were a forerunner of railway tracks rather than wheels.

As we have already seen (p. 4), the *Kao Gong Ji* [*Khao Kung Chi*] chapter of the *Zhou Li* [*Chou Li*] (Record of the Institutions of the Zhou Dynasty) contains a good deal concerning wheelwrights and their work. It is now clear that by the Warring States period (fourth century BC) primitive solid wheels had long given place to very elegantly constructed wheels with hubs, spokes and rims composed of felloes (sectors supported by spokes). The trueness of the wheel was to be such that it resembled a hanging curtain, curving down-wards with beautiful smoothness. In the Han time, elm-wood was used for the hubs, rosewood for the spokes, and oak for the felloes. A curious process of unilateral drying of the wood for the hubs is given. The hub was drilled through to form an empty space into which the tapering axle was fitted, and between the two a tapering bronze bearing was inserted, the exterior being covered by a leather cap to retain lubricant. The thickness of the spokes and the depth of the holes and mortices to receive their tongues and tenons in hub and rim respectively were very carefully regulated, neither too much nor too little, and the legs of the spokes were made thinner towards the rim as a measure of 'streamlining' against deep mud. The number of spokes varied widely. A famous text of perhaps the fourth century BC speaks of chariot-wheels having thirty spokes, and excavations made in 1952 indeed unearthed remains of chariots of this period in which many of the wheels do have this number of spokes. The testing of completed wheels was elaborate, including the use of geometrical instruments, flotation, weighing, and the measuring of the empty spaces in the assembly by millet grains. The interest in these descriptions is perhaps increased when we reflect that though carriage-wheels were those which most interested the scholar-officials, and therefore those which found a prominent place in the *Zhou Li*, the level of Qi and Han crafts-manship revealed is such that gear-wheels (e.g. for water-mills) would clearly have presented little difficulty to the artisans.

It has well been said that the final development of the wheelwright's art was the construction of a wheel not in one plane but as a flattened cone. This

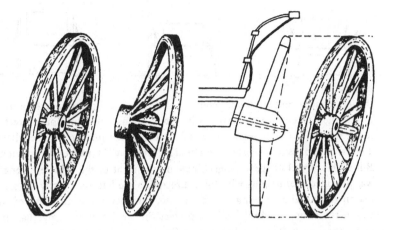

Fig. 264. The dishing of vehicle wheels: drawings from traditional English examples.

is the technique known as 'dishing'. Such wheels give strength against the the sideways thrusts occasioned by the transport of heavy loads over erratic or rutted surfaces. They appear in European illustrations from the fifteenth century AD onwards (Fig. 264). However, Joseph Needham and colleagues have established that dishing, far from being a sixteenth century Western perfection, was systematically employed by the wheelwrights of the late Zhou and Han. This is demonstrated by several passages in the *Kao Gong Ji* as well as later works.

In recent times archaeological excavations have brought abundant confirmation. Investigations of royal tombs at Huixian in northern Henan [Honan] revealed a whole park of nineteen vehicles from the fourth or third century BC. Although the wooden parts had rotted away they had left impressions in the concreted soil so clearly that it was possible to dissect them out down to comparatively small details (Fig. 265). Though quite straight spokes were found, the archaeologists believe they were inserted into the hubs in a slanting manner (Fig. 266).

Very different wheels, however, were found in models of Former Han chariots excavated from tombs of the first century BC at Changsha in Hunan. Here the curvature of the spokes agrees with the *Zhou Li*, but it is so arranged in a shape like a Chinese farmer's hat that the concavity of the dish is inward, not outward (Fig. 267). Still more remarkable, this inward dishing has been perpetuated in traditional Chinese vehicle wheels down to the present time (Fig. 268), though in fact it does not matter much whether the dishing is outwards or inwards. As the figure here shows, the post-Renaissance wheel on

 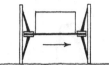 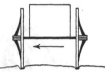

the left, concave outwards and mounted on a downward-pointing axle-bearing on the right of a cart, was particularly strong against jolts with force directed to the right, since these served only to drive the spokes further into the felloes. The same applies to wheels of the Huixian [Hui-hsien] type (centre drawing). But with a wheel of the Changsha type on the right of the vehicle, strength was provided against jolts with force directed to the left, for exactly the same reason (right-hand drawing). In each case, whichever the direction of the strain, the stronger wheel tends to protect the weaker. For an overhanging body, the concavity of the Huixian or European type was more convenient, and also had the advantage of throwing mud clear.

In all ancient and medieval Chinese vehicles, so far as we know, the axle-bearings were horizontal. This was also the case very late in Europe. Such wheels must have had a less enduring life under load than those in which the spokes meet the ground at right-angles, as in the later European types and the very elegant Changsha design. Indeed, some of the Huixian wheels

Fig. 265. Detail of two chariots in the Huixian park, showing how much was recoverable from the compacted soil.

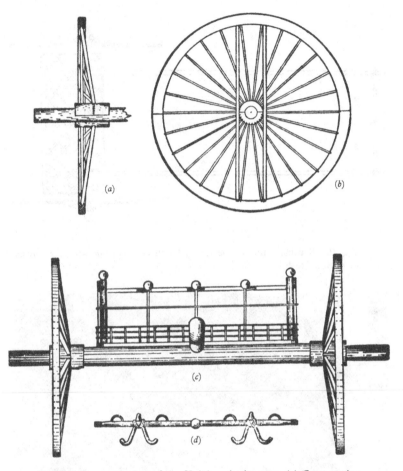

Fig. 266. Reconstruction of the Huixian chariot type. (*a*) Cross-section
of the wheel, showing dish; (*b*) elevation of the wheel, showing
quasi-diametrical struts; (*c*) end-on view of assembled chariot; (*d*) cross-bar
with yokes.

seem almost to admit a tendency to weakness, for they were equipped with
a curious feature about which the *Zhou Li* is quite silent, namely a pair of
struts each close to the diameter of the wheel and passing each side of the
hub (Fig. 266 *b*). These were almost certainly inserted into separate felloes,
thus adding much to the strength of the wheel, holding it dished, and offering
in fact a very early example of a kind of truss construction. One finds no
further evidence of these structures in China itself, but they are to be met

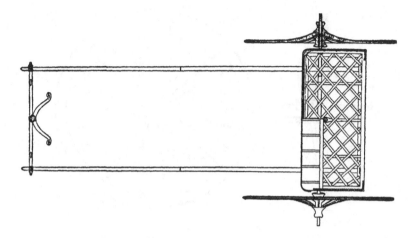

Fig. 267. Reconstruction of the Changsha chariot type, showing the convex or outward dish.

Fig. 268. Two nail-studded cartwheels with outward dish doing duty as flywheels for a manually operated pump.

with today in the country carts of Cambodia, though in a slightly degenerate form (Fig. 269).

Two interesting Han reliefs showing scenes in a wheelwright's workshop have been found by Dr Needham and his colleagues, one of which is reproduced here (Fig. 270). This has a very curious feature, for the partly con-

Fig. 269. Quasi-diametrical struts and outboard axle-bearings on the wheels of a Cambodian farm-cart in 1958.

structed wheel is not easily explicable in terms of text of the *Zhou Li*. Indeed it looks more like a heavy-duty wagon-wheel rather than the fine chariot-wheels there described. Its most unusual feature is that at the end of each spoke there is a block of wood, and as some of the felloes of the rim are already fitted on without overlap, they would seem to be curving plates hiding the blocks from the side, indeed what might be called 'curtain-felloes'. With a counterpart on the other side, the whole would then be covered by strakes of a wooden (or iron) tyre, the intervening places filled in with other blocks, and every component thoroughly pinned together (Fig. 271). This may have given rise to the wheels of the traditional Shandong [Shantung] carts (Fig. 268) in modern times, which are much studded with iron nails clinched at the back over iron washers.

When interest is concentrated upon something which rests on an axle, supported by a suitable framework for the bearings, the wheel is a machine for carrying; but when the chief concern is what lies below it, the wheel is a machine for crushing, as we shall see later. But there is a third function, the wheel as a machine for transmitting rotatory energy. It will be convenient to study this next, because mastery and control of torque has been one of the most fundamental features in the gradual development of machinery.

Fig. 272 is a simple diagram illustrating the variations and combinations of axles and wheels. The simplest form which this may take is the appearance of lugs, often not more than four or six, and often not on a wheel but spaced

Fig. 270. A wheelwright's shop of the Han period; a stone relief from Jiaxiang, near Yanzhou (Shandong), found in 1954. Now in the Jinan Museum. To the left the wheelwright is working on a curved felloe, while another is held by his wife. To the right an assistant is filtering lacquer, paint or glue, with one of the feudal gentry standing behind him.

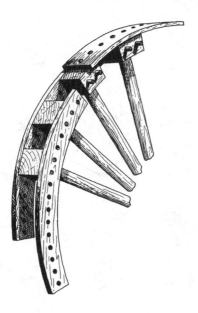

Fig. 271. Conjectural reconstruction of a Han heavy-duty composite cartwheel.

laterally around the shaft itself (Fig. 272 *b'*). When the axle is rotated, the lugs will alternately depress or raise and then release a set of levers or rods conveniently placed – if their heads are weighted we get the mechanical pestle, trip-hammer or stamp-mill. Although there seems to be no evidence of this in Europe before the early twelfth century AD, the trip-hammer was both widespread and ancient in China, and water-power was applied to it already in the Han. It seems clearly one of the simplest methods of utilising rotary motion, easily adaptable for many purposes other than its original use for removing the husk from rice.

The design of the trip-lug can develop, on the one hand, into revolving plates which if non-circular become cams. Though traditional Chinese technology seems to have made little use of continuously rotating cams, by the Han period (about second century BC) it did develop complex cam-shaped rocking lever for the triggers of crossbows. On the other hand, the trip-lug developed into numerous projections, which could take the form of vanes or gear teeth. Vanes are the basis of the water-wheel, either vertical (Fig. 272 *c*) or horizontal (Fig. 272 *c'*). The horizontal water-wheel seems to have been characteristic of the Chinese civilisation from the Han to the Tang, after which the other appears also. We shall see that there is very little difference in the date at which the water-wheel as a power source makes its appearance

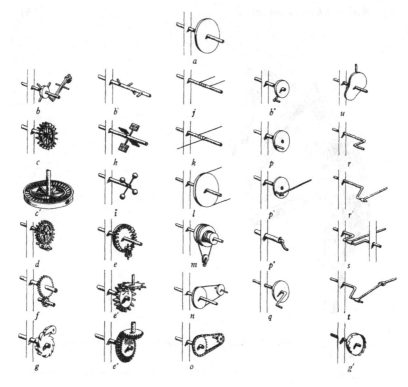

Fig. 272. Variations and combinations of shafts, wheels and cranks.

a wheel, shaft and bearing

b lugs on wheel

b' lugs on shaft

b" handle mounted on lug, forming crank

c vanes in vertical water-wheel

c' vanes in horizontal water-wheel

d flat teeth on two enmeshing gear-wheels

e right-angle gearing; pin-wheel and pin-drum or lantern-wheel

e' right-angle gearing; enmeshing peg teeth

e" right-angle gearing; bevel-wheels

f shaped teeth on two enmeshing gear-wheels

g ratchet-wheel and pawl

g' crown-wheel

h vanes or pedals on spokes

i blocks or bob-weights on spokes

j discontinuous driving-belt; strap-drill or treadle-lathe

k discontinuous driving-belt; bow-drill or pole-lathe

l pulley or windlass

m differential windlass

n continuous or endless driving-belt (with mechanical advantage)

o continuous or endless chain-drive (with mechanical advantage)

p handle at wheel's edge, forming crank

p' handle at wheel's edge, fitted with connecting-rod

p" 'oblique' crank handle

q wheel and crank arm

r crank arm or eccentric lug

r' crank fitted with connecting-rod

s crankshaft and connecting-rod

t crank or eccentric lug, connecting-rod and piston-rod combination, for the interconversion of rotary and longitudinal rectilinear motion

u cam and cam-follower

in China and the West, so that the question of origin is as yet unsolved.

Here we come upon a basic distinction between the direction in which power is transmitted. In using the fall of water as a motive force, the force is exerted on the vanes of the wheel and conveyed to the machinery, but the force may also be applied to the axle of the wheel and conveyed to the water. This last gives us the paddle-wheel boat, to be discussed together with other hydraulic machinery in the next volume of this abridgement.

Passing now from vanes to pegs and teeth (Fig. 272 *d* and *f*), we have before us the history of gear-wheels. In the West these were essentially a Hellenistic development. Their earliest appearance is probably with Ctesibius (*c.* 250 BC) and Philon of Byzantium (*c.* 220 BC) and later with Heron of Alexandria (*c.* AD 60), all of whom used them, or planned the use of them, in a great variety of machines. These men had their contemporaries in China, yet for the Qin and the Former Han (third century BC to the early decades of the first century AD) we have little knowledge of the machines constructed with trains of gears; we can only surmise that much must have been going on since a number of specimens of gear-wheels of that time have survived, and since we find so much textual evidence of the use of machinery with toothed wheels in the Later Han, the Three Kingdoms and the Jin [Chin] Periods (first to fifth centuries AD). They were needed in water-mills, hodometers (distance measuring machines), crossbow-arming mechanisms, south-pointing carriages, chain-pumps and mechanised armillary spheres. Zhang Heng [Chang Hêng] (flourished AD 130), whom we have already met, was famous as a man who could 'make three wheels rotate as if they were one', and his contemporary, Liu Xi [Liu Hsi] (died AD 120), in his dictionary of synonyms, the *Shi Ming* [*Shih Ming*] (Explanation of Names) tells us that the human jaws were familiarly called 'cooperating wheels' or 'toothed wheels', perhaps on the analogy of the intaking tendency of a mill in which two rollers connected by gearing, revolve in opposite senses. Indeed, gear-wheels were so common in the Later Han as to be part of the mental background of official commentators.

Due to the recent expansion of archaeological research in China, there have been many finds of gear-wheels in tombs dating from the Qin onwards through the Han, i.e. from about 230 BC. At least one mould for a bronze toothed wheel has come down to us intact from the Former Han time (Fig. 273 (*a*)); it is made of earthenware. The shank into which the axle would fit is square, and the wheel is a ratchet with sixteen slanting teeth ready to receive a pawl. This is, of course, a special case of gearing needed whenever it is desired that a wheel or roller shall turn only one way and not slip backwards (Fig. 272 *g*). The particular mould in question must have been used about 100 BC for ratchet-wheels forming part of winches, cranes or crossbow arming mechanisms. It is unlikely that the Alexandrians had not also thought of this form of gear, but the earliest Western description of it seems

Fig. 273. Gear-wheels from the Qin period. (*a*) Mould of hard stoneware for a 16-tooth ratchet wheel, *c*. 100 BC; (b) bronze 40-tooth ratchet wheel with large square shank, diam. *c*. 2.5 cm, *c*. 200 BC.

to occur in the work of Oribasius (AD 340 to 400) in connection with surgical instruments.

Another ratchet, this time an actual bronze wheel, was discovered in a tomb full of bronze objects in a village near Yongji [Yung-chi] in Shanxi [Shansi] province. The burial may have occurred in the Warring States period, but is more probably Oin or early Han, *c*. 200 BC. Measuring hardly more than 2.5 cm in diameter (Fig. 273 (*b*)) it has forty teeth and the square shank hole extremely large. The most likely purpose of this ratchet wheel would have been the arming mechanism of a crossbow.

But the most remarkable and unexpected finds have been those pairs of gear-wheels with 'chevron teeth', just like the double helical gears of the twentieth century (Fig. 274). Specimens of these bronze objects were first found in 1953 in a tomb of the early Eastern Han period at a village near Xian [Sian] in Shenxi [Shensi] province, and are dated in the neighbourhood of AD 50. More recently further finds of the same type of gears have come to light from Han tombs at Changsha, further north in the same province. All are certainly Han if not Qin. Still less well known is the fact that there have been a number of finds of iron gear-wheels in Han tombs. One of these, a ratchet wheel of sixteen teeth, is about 7 cm in diameter.

Thus during the three centuries and more preceding the time of Zhang Heng, gear-wheels large and small were being made for a variety of practical purposes. This spread of date is instructive for comparison with the Alexandrian engineers who, broadly speaking, cover the same period, but we

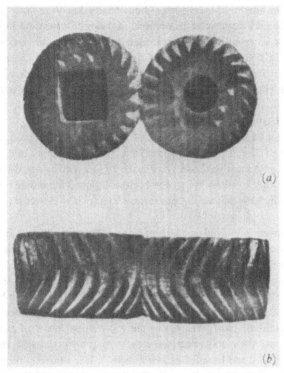

Fig. 274. Objects believed to be Han gear-wheels with chevron teeth (double helical gear). (*a*), (*b*) Bronze 24-tooth enmeshing gears with round and square shanks, diam. 1.5 cm, width 1 cm, *c.* AD 50.

should also notice that the distribution of finds in China is remarkably wide, excluding only the far south and the far north. Interest in gear-wheels and their practical utilisation was therefore not confined to any small region of Han culture. As for the shape of the teeth, all ancient Chinese examples so far seem (if not ratchet-slanted) to approximate to equilateral triangles, exactly as is found in the Greek Anti-Kythera planetary calculator, probably dating from about 87 BC. Rounded teeth occur first in a Chinese hodometer specification of AD 1027 (see p. 187 below), and with an angle of about 30 degrees in Europe around AD 1300. Rounded 'spoke-teeth' are found in the early Western mechanical clocks of the fourteenth and fifteenth centuries AD.

It is fortunate indeed that some of the Han gears were made of bronze and iron, for if all had been made of wood none would have survived to prove knowledge and use. The material employed for them in ancient and medieval China seems to have varied with size and purpose – for water-mills and water-raising machinery they were always certainly of wood, built like wagon-wheels, but for finer machinery such as hodometers they were made of bronze and iron. The only instance of gear-wheels in power-transmission in Western antiquity is a water-mill of Vitruvius. Here a horizontal-axle vertical water-wheel drives a flour-mill by means of right-angle gearing in the form of a pin-wheel and a pin-drum (as in Fig. 272 *e*). The right-angle drive was not necessary in Han China if, as seems likely, the earliest water-mills there were driven by horizontal water-wheels, but it appears abundantly in the Jin (AD 265) and later.

On the other hand, if we knew more of the mechanism adopted about AD 120 by Zhang Heng, and his numerous later successors before the invention of the clock escapement in China early in the eighth century AD, using water-power for rotating armillary spheres and celestial globes, we should very probably have to include this among the earliest examples of power-transmitting gear-work.

After the invention of the escapement in AD 725, gear-wheels flourished in clockwork and its associated moving figures or 'jackwork'. This culminated in the bronze and iron of Su Song's [Su Sung's] elaborate masterpiece of 1088, but it is interesting that, as in the West in the eighteenth century, hardwood for gear-wheels also had its advocates, who appreciated its special lubrication properties and freedom from rust. The pin-drums or lantern-wheels of the Vitruvian water-mill are perhaps the most primitive forms of right-angle gearing – among their successors were the two pin-wheels with enmeshed teeth (Fig. 272 *e'*), and bevelled gear-wheels (*e"*). It is generally thought that the last were quite new when Leonardo sketched them, but four hundred years earlier oblique gears of some kind or other had been prominent in Su Song's clockwork.

From one point of view, gear-wheels consist only of a hub and a number of short spokes. But such spokes can carry objects of different kinds at their ends. From the trip-hammer lugs (Fig. 272 *b, b'*), which were often on the shaft alone, there could be a transition to lugs carrying vanes (*h*) or weights (*i*). In the first case we have the radial pedal or radial treadle, a device of enormous importance in the exploitation of human muscular power, and characteristically Chinese. This was the main 'motor' of the square-pallet chain-pump of the Later Han. This simple form of treadmill was almost unknown in Mediterranean antiquity, and indeed was never adopted by Western peoples. The crank pedal so familiar on bicycles dates only from the middle of the nineteenth century AD. Treadmills in Greek and Roman culture there certainly were, but they were large drums into which one or more men could enter and exert force from within. They persisted through the Middle Ages, but only in AD 1627 are they depicted in a Chinese book. Large and clumsy constructions, they were quite unsuitable for the use of millions of farmers whose problem was to irrigate their land by small machines.

Plates carried on lugs or mounted directly on shafts proved of course to be capable of applications stronger and stranger than the pressure of human feet; they could turn into vanes or sails beaten by the wind – as they first did in early Islamic Persia (Iran) spreading thereafter both east and west (see pp. 271 ff below). This was an 'ex-aerial' development (one in which the air acts upon the device), analogous to that of the water-wheel, but the opposite was equally important for it led to all rotary fans and compressors, 'ad-aerial' devices (where the device acts upon the air) in which the Chinese culture-area took the lead (see pp. 275 ff below). And besides these famous uses, we shall find as we go on certain more unusual functions of blocks fitted on the end of spokes; not only as parts of heavy-duty composite wagon-wheels, but also as carriers for driving belts where an integral rim is absent (pp. 64 and 65).

When radial lugs carry weights, a bob flywheel results. The most ancient was probably the spinning-whorl, a disc-like device that did not find great application before the Paleotechnic period, of which iron and coal were the keynotes. Another example would be the potter's wheel. Second in antiquity would be the heavy discs on bow-drills and pump-drills found in so many cultures. In the eleventh century such flywheels were applied to pestles for grinding. Radial weights (Fig. 272 *i*) found their chief application, perhaps, on the worms of screw-presses from Roman times to the period of printing, and it was doubtless this association with screws, which the Chinese did not possess, which prevented them being much used in China.

The function of a flywheel can also be performed when the balls or bobs are placed, not directly at the end of the spokes, but attached to them by short chains which centrifugal force tautens. This device may be oldest in Tibet, where it has been current for hand-turned prayer-wheels from time immemorial. Such flywheels appear in the West first about AD 1430, but it is by no means fanciful to connect this development with Central Asian domestic slaves who were so numerous in fourteenth- and fifteenth-century Italy, and who may have been responsible for many transmissions of technological interest.

Mention was made above of the use of wheels or rollers for crushing or grinding. In Chinese civilisation these techniques were widespread, and go back almost certainly to the Zhou [Chou]. By combining two rollers, with or without gearing, the cotton-gin and the sugar-cane mill were obtained; these were the ancestors of all steel rolling-mills, mangles, and paper or textile machinery.

This discussion began with rollers, and with them it must also end. For all wheel-carrying shafts must be supported in bearings, and the end of the shaft constitutes a roller. Bearings began no doubt, with the bone or antler hand-holds used by neolithic drillers, and by the second millennium BC Egyptian craftsmen were using bowls of soapstone with which to press on their bow-drills in order to reduce friction. The bearings of the potter's wheel were also of the highest antiquity (the fourth millennium in Mesopotamia); in China small cups of hard porcelain were used for the sockets for a similar purpose, in India small concave flint pebbles. Lubricants are no younger; an Egyptian wall-painting of about 1880 BC shows men transporting a statue on sledges, while one man pours oil or grease along the base of the statue on which he rides.

As we have already seen, the journals and bearings of vehicle axles were of bronze or iron in Zhou and Han China just as they were in Roman antiquity. However, it is sometimes said that after the fall of the Roman empire there appear no more metal bearing-surfaces until the fourteenth century AD; this may possibly be true of Europe but is certainly wrong for China. The mechanised armillary spheres between the second and eighth centuries AD could not have worked at all without metal bearings. About AD 720 we even hear of steel bearings being used for this purpose, and when we come to the clock tower of Su Song in 1088 the details are particularly clear.

Although in these remarkable machines we have not encountered the use of intermediary rolling objects for taking off friction, China may be deeply involved in the pre-history of ball-bearings. After the stonemasons of high antiquity we meet first with a systematic use of rollers in the battering ram and the gate-borer invented by Diades, one of the engineers of Alexander the Great, whom he accompanied on his campaigns (334 to 323 BC). But here the rollers are still in a straight line, not arranged peripherally around a shaft. For the earliest roller-bearings we may have to look to China. Among the finds at a village in Shanxi in the late 1950s, there were some remarkable circularly shaped bronze objects with internal grooves. These channels were divided into four or eight compartments by small partitions, and each of these was filled with a mass of iron dust grains. Several of these possible 'ball-races' or roller bearings, of different sizes, were found. Since the objects of this tomb must be dated at least as early as the second century BC, it would seem that the Chinese were taking an early interest in the free and smooth running of shafts and axles.

If the rust came from balls or rollers, these objects would certainly be the oldest ball-bearings known. If it did not, pride of place must still go to the strange trunnion bearings of the capstans of the Roman ships built between AD 44 and 54, and recovered this century from the Lake of Nemi south of Rome. Strictly speaking, the latter were roller-bearings and not true ball-bearings, because the balls could rotate in only one plane, but they are of great ancestral interest. Trunnions recur in Islam in AD 1206, but as we shall see later, certain imperial vehicles constructed in China between the seventh and eleventh centuries showed a steadiness hard to account for otherwise.

PULLEYS, DRIVING-BELTS AND CHAIN-DRIVES

We now return to the point of departure already mentioned, namely the investigation of what happens when fibrous bodies (sinews, thongs, cords, ropes, etc.), or chains, are wrapped round rotating axles and wheels. Fig. 272 *j* shows the simple arrangement of the strap-drill (p. 31 above) or treadle-lathe, and *k* that of the bow-drill or pole-lathe (p. 32 above) – all methods which produce only alternating rotary motion. To have it continuous, the endless driving-belt is necessary (*n*), and this was therefore an invention of first-rate importance. Before considering it, however, a few words must be said about the simple pulley (*l*), the wheel with a furrowed rim able to retain a rope or cord running over it. Who first realised that friction of a rope passing over a projection would be enormously reduced if a wheel was interposed is unknown, but the practice was familiar both in Babylonia and ancient Egypt.

Pulleys (*lu-lu* or *lo-lu*) would therefore be expected in China from very early times. Fig. 275 shows a famous relief from the Wuliang tomb-shrines (AD 147) depicting the recovery of Zhou [Chou] cauldrons from the river, where two crane pulleys may be inferred. But the *Li Ji* [*Li Chi*] (Record of Rites) of AD 80 to 105, which contains some earlier texts, reports a story about Gongshu Pan [Kungshu Phan] (fifth century BC) and the tackle used for lowering the heavy coffin-lids at important burials. And according to commentators and Liu Xi's [Liu Hsi's] *Shi Ming* [*Shih Ming*] (Explanation of Names), the term *feng bei* [*fêng pei*] seems to have meant a kind of four-posted derrick with four pulleys. In this connection it is interesting that the earliest (Han) pictures of the derricks over the Sichuanese [Szechuanese] brine wells also show four-posted structures, not pyramidal. Indeed, pulleys were so common in the Han that other things were named after them, such as swords with 'pulley-wheel' shaped hand-guards. Pulleys were also in constant demand for palace entertainments – for example, about AD 915 a whole *corps de ballet* of two hundred and twenty girls was hauled up a sloping way from a lake in boats.

These heavier uses approximate the pulley to the winch, windlass or capstan. The ordinary windlass is shown in nearly all the Chinese books which illustrate machines, beginning with AD 1313. The main interest in so ancient an instrument is the extent to which it incorporated the crank principle, and to this we shall shortly return. The winch was of course also used for tautening when mounted with a ratchet.

The history of the related machine known as the 'Chinese windlass' is also obscure. Old Western textbooks of physics sometimes give it as an example of differential motion, for the axle carries two drums of different diameter (Fig. 276) on which the rope winds and unwinds continuously, thus giving useful work and amplification of effort. No one has yet succeeded in finding any Chinese text which would justify the name which the device bears in the

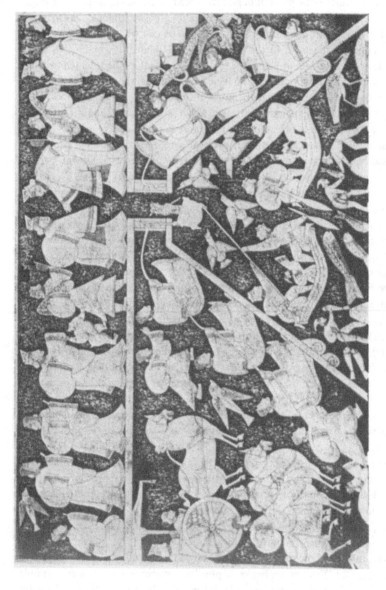

Fig. 275. Crane pulleys in the Wuliang tomb-shrine reliefs. Illustrated is the unsuccessful attempt by Qin Shihuangdi to recover the cauldrons of the Zhou.

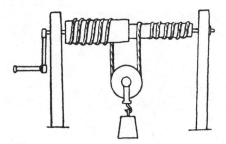

Fig. 276. The differential, or 'Chinese' windlass.

West. Possibly it is due to a misunderstanding, since Chinese windlasses (e.g. in coal mines, gem mines or wells) commonly have two coils of rope on the same drum, one bucket or seat being lowered as the other comes up. The idea of the compound drum might of course have been derived from this, and the invention may well have been a local one in China which was never described in literature.

A minor invention of the windlass type which is in all probability Chinese is that of the reel on a fishing-rod. The oldest known painting of this is by Ma Yuan, who flourished towards the end of the twelfth century AD (Fig. 277). The advanced character of the Chinese textile industry, with its numerous bobbins and reels, may be connected with the invention. However, there is a possible very early reference, the story of Lingyang Ziming [Ling-yang Tzu-Ming] in *Lie Xian Zhuan* [*Lieh Hsien Chuan*] (Lives of the Famous Xian (Immortals)). In the text, which may be dated as of the third or fourth century AD, the Daoist fisherman Lingyang Ziming, who once caught a white dragon with his fishing-rod, was rapt away by it to live the life of an immortal on the holy Lingyang mountain. Years later, a Daoist disciple came and asked the country people whether the *diao che* [*tiao chhê*] (鹿車) of Lingyang Ziming still existed – intending, no doubt, to try his fisherman's luck in turn. All that matters for us is the expression itself, and it can hardly mean anything other than the reel of the rod.

Far more important is the continuous driving-belt, Fig. 272 *n*. Here at the outset it is necessary to make a distinction between belts or cords transmitting power, and similar endless belts conveying material. In the West the second of these uses seems to have preceded the first. A parallel distinction has to be made for the more sophisticated and more efficient chain-drive worked by sprocket-wheels and thus overcoming the demon of slip. It will be convenient to deal first with belts and then with chains.

Strangely there seems no evidence for the driving-belt in Graeco-Roman antiquity. In its purely conveyor form, it may have been present in some of

Fig. 277. 'Angler on a Wintry Lake'; the earliest illustration of the fishing rod reel, a painting by Ma Yuan, *c.* AD 1195.

the earliest examples of the 'chain of pots', the water-raising device known as the *sāqīya*, if any of these were ever made with a wheel at the lower end of the band's excursion. But this was no power transmitter. The earliest picture of a driving-belt which European historians of science have been able to bring forward is that of a rotary horizontal grindstone in the manuscript dating from AD 1430 of an anonymous Hussite engineer. Even in the seventeenth and eighteenth centuries, representations of driving-belts remain rare.

If Asia is to be accorded greater credit for the development of the driving-belt, this may well turn out to be due to the origin there of the only really long-staple textile fibre of antiquity, silk. The spinning-wheel, in its various forms, with its continuous rotary motion assured by a driving-belt, was indispensable for reeling such a fibre. At first sight this is an integral part of the technology of short-staple fibres. Naturally enough, its prominence in our own Western culture has obscured the fact that it had precursors or predecessors in quite another part of the world, designed to handle a totally different sort of fibre, for which continuous rotary motion was more important still.

Fig. 278. The Chinese multiple-spindle spinning machine, an illustration of
AD 1313. The three spindles are all rotated by one continuous driving-belt.

Though there has been great uncertainty about the origins of the spinning-
wheel, Chinese textile technology now appears to be the focus of origin of
all such belt-drive machinery. The matter is of great moment, for the
spinning-wheel embodied not only power-transmission by 'wrapping connection' but also one of the earliest uses of the flywheel principle as well.

That the spinning-wheel appears in Europe very late has long been known.
While drawings of it in the fifteenth century AD are numerous, the oldest
datable Western illustration of it is usually taken to be that in the Luttrell
Psalter of about AD 1338. But Chinese illustrations take precedence. The
multiple-spindle cord-making machine of Leonardo is almost an exact copy
of the Chinese multiple-spindle spinning machines illustrated from 1313
onwards. These are usually shown with driving-wheels having integral rims
(Fig. 278). This drawing also shows a treadle, but one connected to the wheel
itself at one end. The presence of three and even five spindles, all driven by

Fig. 279. The rimless driving-wheel, a form in which the ends of the spokes are connected by thin cords to form a 'cat's cradle' on which the driving-belt is carried. Cotton spinners at a cave-dwelling in Shenxi, *c*. 1942.

one belt in these machines, early in the fourteenth century, seems to give them a stamp of maturity, and suggests that at that time they had already long been developed. Rimless driving-wheels are also common in Chinese textile technology. In one type the outwardly diverging spokes are connected by thin cords so as to form a bed or 'cat's cradle' which carries the driving-belt, while in others it passes over grooved blocks set at the end of the spokes. An apparatus of the former type is being used by the old soldier in Fig. 279 sitting outside a Shanxi cave-dwelling during the Second World War. Interestingly the machine is identical in every respect with that in the picture which is a very early representation of a spinning-wheel (Fig. 280), part of a painting attributed to Qian Xuan [Chhien Hsüan] and in any case datable about AD 1270; a dutiful son is taking leave of his mother, who turns the crank handle with her right hand and spins the yarn with her left. The second arrangement is seen in Fig. 281, a photograph taken by Dr Needham at Dunhuang in Ganzhou 1943, where this kind of spinning-wheel was used for twisting the threads of hemp or heavy flax, or for making thick grain-sack yarn from the scrapings of hides. The construction here invites comparison with the curious type of composite wheel built in the Han (Fig. 271), and again raises the problem of a connection between the construction of vehicular and stationary wheels. Figures 278 and 281 display a prominent crescent-shaped mounting for the spindles in machines where the driving-belt is single. But this disappears when multiple driving-cords are used, one for each spindle, as used in a hand-crank spinning-wheel (Fig. 282) and in silk-throwing machinery.

To sum up, Chinese illustrations of the spinning-wheel have clear priority over those of Europe. During the thirteenth century AD, cotton culture was spreading over China for the second time, probably from Xinjiang [Sinkiang], but this may not have been any limiting factor for from quite ancient times the Chinese had been using fibres other than silk which needed spinning, notably hemp and ramie. The spinning-wheel may thus have originated at any time after the Han.

Actually, it is not necessary to go outside silk technology to look for the birth of the spinning-wheel. The Chinese never wasted anything, thus they early developed means of dealing with waste silk on cocoons from which the moths had escaped, and coarse silk which could not be wound off from cocoons in the classical way. In fact, if the beginnings of sericulture were concerned with 'wild' silkworms, as we must surely assume, this problem would have been the oldest of all. In AD 1313 Wang Zhen [Wang Chên] in his *Nong Shu* [*Nung Shu*] (Treatise on Agriculture) shows us the making of floss silk for wadding, etc., from waste cocoons agitated in boiling water; the best of this goes, he says, to make *mian* [*mien*] (綿) a waste silk which could be spun, the coarsest to make *xu* [*hsü*] (絮) or wadding. On the very next page, he shows us a woman combining together the longer strands end to end by hand

Fig. 280. An early representation of the spinning-wheel; a painting attributed to Qian Xuan and in any case datable about AD 1270. A son is saying farewell to his mother, who turns the crank handle with her right hand and spins the yarn with her left.

Fig. 281. The block-spoked driving-wheel, a form in which the ends of the spokes bear grooved blocks on which the driving-belt is carried. Here the five spindles of the crescent-shaped mounting are used for spinning hemp, flax, etc.

with the aid of a 'twisting spindle' – in fact spinning: the production of yarn for rough silk fabric. Although he does not depict any kind of spinning-wheel in use for this purpose, he says that what she is doing is a substitute for it. And then when we turn to contemporary descriptions of the Chinese silk industry we find at once that true spinning-wheels were and are in fact traditionally used for making silk yarn from short lengths directly from the wild cocoons themselves. Thus we may well have here the real point of origin of all spinning-wheels.

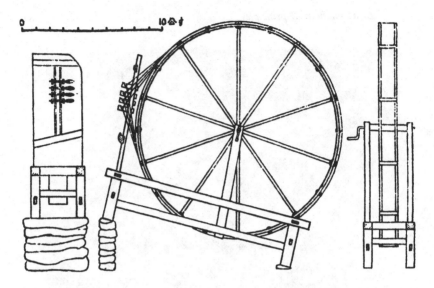

Fig. 282. Scale drawing of a Sichuanese spinning- or quilling-wheel with four spindles each rotated by a separate driving-cord from the main wheel. With multiple driving-cords no crescent-shaped mounting for the spindles is necessary.

It has been suggested that the immediate precursor of the Western spinning-wheel was the quilling-wheel for winding yarn on the bobbins of the weavers' shuttles. This also had its driving-wheel, and a small pulley, mounted on bearings in a framework, and connected by an endless belt. Yet Chinese pictures and references again long predate it. In Europe the famous windows of Chartres cathedral (AD 1240 to 1245) depict quilling-machines, but there is no earlier sign. As to the other end of the Old World, we can go back to the *Geng Zhi Tu* [*Kêng Chih Thu*] (Pictures of Tilling and Weaving) compiled by Lou Shou about AD 1145. The later editions do not show a quilling-wheel under 'Weft', but by good fortune that of AD 1237 clearly does, and at its right-hand side we find the original poem of Lou Shou, which runs as follows:

Steeping the wet again*, is part of the mystery,
And with hands as cool as the bamboo shoots of spring,
The country maidens marry two fibres of silk
And twist them together to one inseparable thread
Fitted to play its part in a myriad patterns.
And now, at last, in the late slanting radiance,
The big wheel's shadow looks like the toad in the moon.**
Yet still, here below, sweet Ah Xiang speeds her turning,
And under a deep blue vault the rumbling goes on.***

*This refers to the de-gumming process;
**The wheel in this line is clearly the driving-wheel of the quilling or throwing machine. Its outstretched spokes without a solid rim make it look like a spread-eagled toad.
***There is a significance in the noise in the onomatopoeic Chinese words *li-lu, lu-lu, lo-lu,* etc. connected with the creaking and squealing of primitive bearings.

Thus we need have no hesitation in accepting the use of the quilling-wheel for the middle of the twelfth century AD in China, nearly a hundred years before the first European appearance. More remarkable still, texts and reliefs dating from the first century BC to the third century AD, consideration of which must be postponed until later (p. 174) when we discuss the Chinese wheelbarrow, take the story of the quilling-wheel fully back to the beginning of our era. Such a priority can only be explained by the stimulus of an industry which had to deal with extremely long continuous fibres.

Indeed, if the belt-drive developed so long ago for winding shuttle-bobbins and for silk-throwing, it is highly probable that it was applied quite early to the most fundamental of all sericultural operations, the reeling or winding-off of the silk filaments from unbroken cocoons. In this connection, one can reconstruct completely the classical silk-reeling machine with its oscillating 'proto-flyer' from a text of approximately AD 1090, the *Can Shu* [*Tshan Shu*] (Book of Sericulture) of Qin Guan [Chhin Kuan], by the aid of 17th century AD illustrations (Fig. 283). In this apparatus the main reel on which the silk is wound is powered by a treadle motion, and the ramping-arm of the flyer (a device for spreading the threads evenly across the reel), is activated simultaneously by a subsidiary belt-drive. The machine is clearly recognisable in the AD 1237 drawing in the *Geng Zhi Tu* [*Kêng Chih Thu*] (Pictures of Tilling and Weaving). The construction is so important for the history of the driving-belt that readers may like to have the actual words of the *Can Shu*. It says: 'The pulley (bearing the eccentric lug) is provided with a groove for the reception of the driving-belt, an endless band which responds to the movement of the machine by continuously rotating the pulley.'

Although in this apparatus the belt-drive (thus attested for the eleventh century AD) is subsidiary to the main motion, while in the quilling-wheel or spinning-wheel it is essential to it, the machine itself is perhaps more fundamental than either of them. It may be, therefore, that the most ancient use of the power-transmitting endless belt was for the purpose of laying down the fresh silk filaments evenly on the reel. The fact that in the reeling-machine the transmission of power involves no mechanical advantage might also invite us to think it older. In any case it would be rather surprising if the reeling-machine were not at least as old as the quilling-wheel.

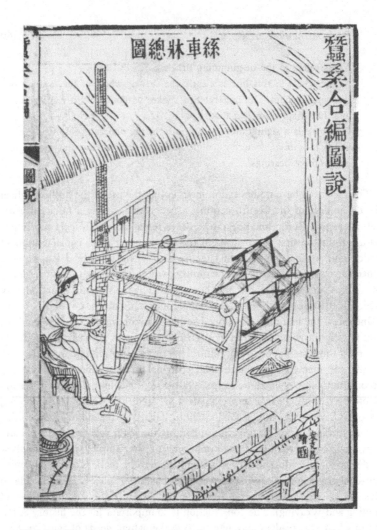

Fig. 283. The classical Chinese silk-reeling machine, an illustration from the
Can Sang He Bian of 1843. Although this treatise on silk is so late, it is
strictly traditional in character, closely following seventeenth-century
models, and corresponding in every detail of its mechanical descriptions
with the literary evidence the late eleventh-century *Can Shu* of Qin Guan,
which it faithfully illustrates. The individual fibres of silk are drawn from
cocoons in the heated bath, passing through guiding eyes and over rollers
before being laid down on the main reel. All the motions of the machine
originate from a single treadle action of the operator. This rotates the main
reel by means of a crank, but at the same time also a pulley (at the other
end of the frame) which is fitted with an eccentric lug, makes the
ramping-arm move regularly. The power is transmitted by a driving-belt.
This machine is important in the history of technology in several ways.
The inscription at the bottom right-hand corner says 'Drawn by Yuan
Gezhang [Yuan Kechang]'.

Textile technology is not the only field which gives us driving-belts in medieval China. The endless rope or cord may go back to the first century AD with the metallurgical bellows, though this will depend also upon the nature of the mechanism driving the bellows.

Thus far we have been considering 'wrapping connection' as applied to smooth-treaded wheels of the pulley type. But toothed wheels may also participate in this principle, and when gear-teeth were made to enmesh with the hollow spaces in the links of a chain, a new device of great power and potential precision was added to the engineer's stock-in-trade.

The history of chain-drives (Fig. 272 *o*) is slightly different, as they seem to have been appreciated a little better in Western antiquity than driving-belts. Philon, about 200 BC, described the endless chain of pots for raising water so characteristic of later Arabic culture that they bear an Arabic name *sāqīya*, and so did Vitruvius (30 BC) after him, but in this type of chain-pump no power was transmitted.

Philon of Byzantium is also associated with another use of the endless chain, in his repeating catapult. Here there were two endless chains, one each side of the stock held for firing, to claw back the string of a torsion catapult, but there is great doubt whether it was used on any extensive scale, or even constructed at all. At all events, the chains did not transmit power from one shaft to another. In this sense the chain-drive seems to have come with much delay in Europe, arriving only in the eighteenth or nineteenth century. However, about AD 1438 Jacopo Mariano Taccola figured an endless chain for hand use – like those employed in operating small hoists in engineering workshops today. But not until 1770 did Jacques de Vaucanson develop an industrially practical chain-drive, which he used in his silk reeling and throwing mills. Hinged links to mate with sprockets on wheels arrived in 1832, after which the chain was applied to cars in 1863 and to bicycles in 1869.

In China development took place quite otherwise. The most characteristic water-raising machine, the radial-treadle square-pallet chain-pump, must necessarily have a sprocket-wheel at each end of the pump's inclined channel or flume up which the water passes. Its invention lay in the first century AD during the Han, yet it was no more a power-transmitting machine than the chain-pumps of the Roman Empire.

But it probably inspired the true invention. When we study later on (chapter 5) the work of the builders of mechanical clocks in China at the beginning of the eighth century AD, we shall find that considerable use was made of real chain-drives – 'celestial ladders' (*tian ti* [*thien thi*]) (天梯). In the great astronomical clock tower of AD 1090 the main vertical transmission-shaft proved to be too long, and was soon replaced by modifications in which the power for the armillary sphere on the upper platform was provided by endless chain-drives, successively shorter and therefore more efficient in the various rebuildings. In this masterpiece of Su Song's [Su Sung's], the uppermost shaft

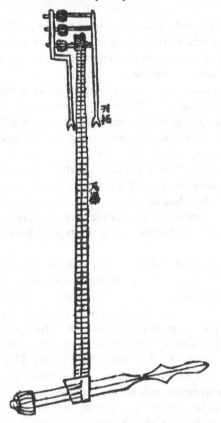

Fig. 284. The oldest known illustration of an endless power-transmitting chain-drive, from Su Song's *Xin Yi Xiang Fa Yao* (New Design for an Astronomical Clock) of AD 1090. This 'celestial ladder' was used in the first and second modifications of the great astronomical clock tower at Kaifeng for coupling the main driving-shaft to the gearing for the armillary sphere (see Fig. 372 below).

rotated a series of three small pinions in a gear-box underneath the armillary sphere at the top of the edifice (Fig. 284). But this was by no means the earliest, for there is some reason to believe that the mercury-operated clock built by Zhang Sixun [Chang Ssu-Hsün] in AD 978 also contained a chain-drive. Thus it would seem that China rather than Europe is responsible for both of the fundamental inventions of the driving-belt and the chain-drive.

CRANK AND ECCENTRIC MOTION

Of all mechanical discoveries that of the crank (*qu huai* [*chhü huai*]) (曲拐) is perhaps highest in importance, since it permits the simplest interconversion of rotary and reciprocating (i.e. straight line) motion. 'Continuous rotary motion', wrote the historian Lynn White, 'is typical of inorganic matter, while reciprocating motion is the sole form of movement found in living things. The crank connects these two kinds of motion; therefore we who are organic find that crank motion does not come easily to us . . . To use a crank, our muscles and tendons must relate themselves to the motion of galaxies and electrons. From this inhuman adventure our race long recoiled.' The principle forms of the device are shown in Fig. 272. The simplest, and perhaps the oldest (or more probably the second oldest), manifestation of the device was born when it occurred to someone that a wheel could be turned by hand more easily if a handle at right-angles to its plane were fixed in it at a point near its circumference (Fig. 272 *p*). In due course the direct contact of the worker's arm was replaced by a connecting-rod (shown in *p'*). Greater leverage was obtained when the handle was mounted on a lug or spoke extending from the wheel's rim, as at (*b''*), and of course this system may have itself have been a modification of the capstan or windlass handspikes (*b*). As a substitute (whether primitive or degenerative one can hardly say) a piece of wood inserted in the axle at an angle (*p''*) was found to be capable of performing the office of a crank. Then the crank proper, invisible in (*p*) and rudimentary in (*b''*), could manifest itself fully in the developed crank arm or handle as at (*q*). Conversely, the wheel could also be invisible (see p. 56 above), leaving only the crank arm or eccentric lug, (*r*) which could carry a connecting-rod (*r'*). Much greater rigidity was acquired by the machine when this was doubled to form a crankshaft (*s*), and the conversion of rotary to perfectly rectilinear motion, as of a piston, was obtained when the connecting-rod was joined by a link to a second rod as in (*t*).

Many historians of technology have been of the belief that the crank appeared rather late. Though cranks appear in illustrations reconstructing the devices of Heron and others in Alexandria, there is nothing in their texts to support this. The first evidence in Europe for a crank-handle appears to be in the Utrecht Psalter, written about AD 830, where it is seen being used with a rotary grindstone. As far as the crankshaft is concerned, this does not appear until the 14th century AD.

It is not difficult to see how the compound crank or crankshaft was born. The East Anglian Luttrell Psalter of about AD 1338 has a rotary grindstone worked by two men using crank-handles placed 180 degrees apart. To originate the form of the crankshaft it was thus only necessary to place two crank-handles in line at the ends of a drum; but mechanical genius must have been required to see the advantages of transmitting torque to or from the centre by embodying, as it were, the eccentric in the shaft, not only at its two ends. If Guido da Vigevano (*c.* 1280–1350) was not the first to appreciate this, he was of the first generation. But there is more to say about the origin of the basic crank itself.

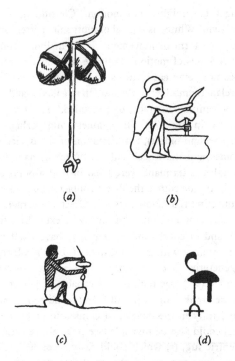

Fig. 285. Ancient Egyptian crank drills, forerunners of the brace-and-bit drill, from Old Kingdom reliefs. (*a*) Oblique-crank flint drill with stone weights; (*b*) a similar tool in use for drilling out a stone pot; (*c*) drilling out a vase; (*d*) Middle Kingdom hieroglyph for the crank drill.

It has long been known that what may have been a crank drill of primitive brace-and-bit type, goes back to the Old Kingdom of ancient Egypt (about 2600 to 2150 BC), (Fig. 285). It even possessed a special hieroglyph (Fig. 285 *d*), which seems to depict the brace and bit with a handle more crank-like than it really was, assuming no development had occurred during the intervening centuries. The tool seems to have spread very slowly out of Egypt; it may have reached Assyria but seems to have been unknown in ancient Greece or Rome. The earliest European illustrations of the crank drill appear in German, French and Flemish paintings dating from the neighbourhood of AD 1420 onwards. Although the tool was now a crankshaft rather than a crank, the connecting-rod component was of course still the worker's arm, but so also it had been in the case of the ordinary grain mill or quern. The first origin of the crank may still have to be sought long before the beginning of our era.

The obliquely pointing position which the handles of some of the ancient Egyptian crank-drills seem to have had is interesting. Assuming for the sake of argument that this almost diagonal substitute for two right angles was a primitive feature, it is striking that it has persisted to the present day in some of the rough windlasses of Chinese miners and farmers. It occurs also on well-windlasses in tomb-paintings, notably some of the Jin dynasty (twelfth century AD) in Shanxi. The oblique 'crank' is also found on some early European hand-querns of about the first century BC.

The most extraordinary kind of oblique crank known to Dr Needham is the treadle crank found on some Chinese spinning-wheels (Fig. 278) and peculiar to them. In this the treadle is attached to one of the spokes of the wheel nearer the hub than the rim, while its other end is held loosely in the framework; one can see that if it were to be copied in an accurate material such as metal, at least one universal joint would be required. This remarkable device seems to be purely Chinese. But it has a certain analogy in the long crank-handles of some medieval European hand-mills.

As to the common cranked handle used to hoist and lower buckets in wells, the earliest evidence for its use in Europe seems to be from AD 1425, so Chinese medieval cranked well-hoists therefore take precedence, and not only for the oblique variety but also for the regular right-angled variety. The oldest illustration is in the *Nong Shu* [*Nung Shu*] (Treatise on Agriculture), implying therefore the thirteenth century AD, and there can be no doubt about it because the construction is clearly described in the accompanying text.

The simple quern (see Fig. 301 p. 115) with its upstanding handle near the rim of the mill-stone has long been recognised as embodying the principle of the crank. Another type (see Fig. 303 p. 118), giving greater mechanical advantage or leverage by increasing the eccentricity, demonstrates Fig. 272 (*b″*). This constitutes, as we saw, a transitional form between the handspikes of the capstan or winch and the crank. Finally, this type also acquired a connecting-rod of varying length (Fig. 286). In so doing, it gave birth to the eccentric lug system, which was great favourite with the medieval Chinese engineers. One finds it, for example, in the fourteenth century AD version of the water-driven metallurgical bellows of the first century AD, where it turned at higher speed on a small pulley connected with a large driving-wheel by a belt. One finds it also on the silk-reeling machine of the eleventh century AD (above, p. 70), where it operated an early type of flyer for laying down the silk in even layers on the reel, and again was rotated by a belt-drive. The textile apparatus may well have been a direct precursor of the hydraulic blowing-engine, the two being safely dated to the ninth and twelfth centuries AD respectively, and the latter is a machine of great importance in the history of technology for it embodies, as far as Dr Needham can see, the first appearance of that fundamental combination of eccentric, connecting-rod and piston-rod (Fig. 272 *t*) used afterwards in all steam and internal combustion engines. He does not know, however, of any Chinese machine which involved the crankshaft principle (Fig. 272 *s*); indeed, it does not appear to be illustrated

Fig. 286. Crank in the form of an eccentric lug, fitted with connecting-rod and hand-bar to allow for the simultaneous labour of several persons at a mill for removing the husks from grain.

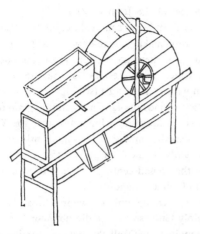

Fig. 287. The Chinese rotary-fan winnowing-machine in its classical form, from a drawing of AD 1313.

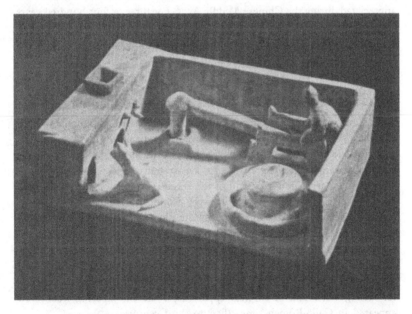

Fig. 288. Farmyard model in iridescent green glazed pottery from a tomb of the Han period (second century BC to second century AD) in the Nelson Art Gallery, Atkins Museum, Kansas City. On the right a rotary grain mill and a man working a foot tilt-hammer for pounding grain; on the left a built-in winnowing machine with hopper and two lower apertures. The crank-handle to the right of the hopper is for working the rotary fan of the winnowing-machine and is the oldest representation of the true crank-handle from any civilisation.

in a Chinese book before the coming of the Jesuits (AD 1627).

Crank-handles were, of course, far from being confined to querns in China. They were used in machines for rope-making, wire-drawing, silk-winding, etc., but the question is how far back these go. Certain museums possess Chinese tomb models in pottery which show, besides the usual quern and foot tilt-hammer, a winnowing-fan with crank-handle (see Figs. 287 and 288). Now there is no doubt that the rotary winnowing-fan for separating grain from chaff, is a machine of considerable antiquity in China, and that the West received it from there no earlier than the middle of the eighteenth century AD. The model of Fig. 288 is certainly of the Han period; that is to say it must date from before the end of the second century AD. It is to be saluted as embodying the most ancient indubitable hand-crank.

It is now certain that crank-handles were not only known in China, but also applied to many machines. Certainly Chinese culture did not spontaneously generate modern science and technology, with all the industrialisation that developing capitalism in the West drew forth from it, but within the limits of a feudal-bureaucratic society the powers of the crank were widely used and appreciated during the Chinese Middle Ages. For three or four hundred years before the time of Marco Polo (AD 1254–1324) it was employed in agricultural and textile machinery, in metallurgy for blowing-engines, and in humbler uses such as the windlass.

SCREWS, WORMS, AND VANES ARRANGED IN HELIXES

The continuously winding screw-thread, male and female (as in bolt and nut), and the cylindrical worm capable of engaging with an ordinary gear-wheel so that motion may be transmitted between two shafts at right angles, are the most outstanding examples of mechanical systems apparently unknown to Chinese engineers and artisans until the seventeenth century AD. Yet in the West the principle of the screw was very familiar in Hellenistic times; the reputed inventor was Archytas of Tarentum (flourished 365 BC). Particularly common were the worm- or screw-presses used in the wine and oil industries, shown, for instance, in the wall-paintings in Pompeii. Many medieval examples still exist in Europe, but China always used wedge presses. In the West there were other notable uses of worms, such as the Archimedean screw for raising water and in surgical apparatus, while the tapering woodscrew appeared in Gallo-Roman times (first to third centuries AD).

In Chinese literature the first picture of a screw occurs in the *San Cai Tu Hui* [*San Tshai Thu Hui*] (Universal Encyclopaedia) of AD 1609 where it is shown used for the cap of a gun. It was also at the beginning of the seventeenth century that the Archimedean screw for raising water was introduced into China, when it was repeatedly illustrated in books. But there is little evidence that it ever spread there, or replaced to any appreciable extent the traditional square-pallet chain-pump.

It seems almost certain that worm gearing reached the frontiers of the Chinese culture-area with the arrival of cotton. For in the cotton-gin of India, a simple machine for separating the seeds from the cotton itself, two oppositely rotating rollers are made to engage not by ordinary gear-wheels but by elongated worms placed side by side. The whole can thus be worked by one crank-handle. Since cotton technology is indigenous to India and very ancient there, the remarkable gearing in these devices raises the question whether the worm (and hence the screw) did not originate in the first place in India rather than Greece. It is curious that the principle of parallel worms found no use in Europe until the early nineteenth century, and the modern use of worms is also in general quite different, for right-angle gearing with speed reduction.

In China cotton-gins had no gearing. The *Nong Shu* of 1313 provides its earliest illustration (Fig. 289) and shows the machine having two crank-handles. As it required two operators an improvement took place in course of time whereby only the lower roller was worked by a hand-crank, while the upper one (smaller and of iron) was worked by a treadle motion assisted by a radial flywheel (Fig. 290).

From this it would seem therefore that in entering China the cotton-gin left its worm-gear behind. But it does not quite follow that this was a case of refusal to accept a piece of technique foreign to indigenous practices; there may have been machinery already 'in occupation'. For while true cotton was native to India and had been known almost since the Han as *ji-bei* [*chi-pei*]

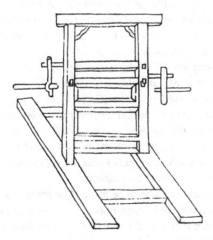

Fig. 289. A Chinese cotton-gin of AD 1313. Each roller has a crank-handle.

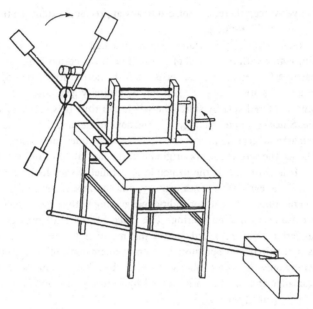

Fig. 290. Another type of Chinese cotton-gin, in which one roller is rotated by a crank-handle while the other is worked by a foot treadle with the aid of an eccentric on a bob flywheel.

(吉貝) as well as by other names, the fibres from the silk-cotton or kapok tree (*mu mian* [*mu mien*] (木棉)) had been used as a textile in China from an early date, principally by the tribal Man peoples of the south. In the Song [Sung] and Yuan, the phrase *mu mian* was carried over as a name for true cotton, as may be seen indeed in the *Nong Sang Ji Yao* [*Nung Sang Chi Yao*] (Fundamentals of Agriculture and Sericulture) of AD 1275.

True cotton seems to have come into China by two routes; through Burma and Indo-China in the sixth century AD, and through Xinjiang [Sinkiang] in the thirteenth. Thus worm-gear would have been offered twice. But it may well be that the double powered rollers of the Chinese machines were already at work from the Han onwards for the *mu mian*, and naturally continued in use when true cotton largely supplanted it, becoming indeed one of China's chief textile fibres. In any case the whole story shows once again that the screw principle was not characteristic of Chinese technology.

Perhaps the most interesting point remains to be made. The absence of the continuous screw and worm in traditional China must not be taken to mean that no forms of helix were known there; on the contrary, some of these were quite venerable inhabitants. For example, there was the zoetrope or vane-wheel rotated by an ascending air-current (see pp. 361 ff in volume 2 of this abridgement), and the helicopter top or horizontal airscrew which itself ascended when rapidly rotated by a cord (see p. 282 below). The skew-set vanes of all such devices were essentially separate flat surfaces tangent to the continuously curving helix of a complete screw or worm. It is thus possible to define rather precisely the different achievements of the Hellenistic and Chinese worlds, for while the former made abundant applications of elongated screw and worm shapes, the latter early developed tangent plane helicoid structures.

In the late fifteenth century AD European engineers were placing vane-wheels in kitchen chimneys with shafts and gearing or turning spits; Leonardo himself designed one of these. It seems likely that this use of ascending hot-air currents derived from the earlier zoetropes of China (which go back to the Tang if not to the Han) and the prayer-wheels of Mongolia and Tibet. The transmission could have occurred very easily through the domestic slaves from Central Asia who were brought to Italy in great numbers during the four-teenth and fifteenth centuries AD. As the zoetrope-spits became so common in the great houses of Europe, it seems equally likely that they played a part alongside Hellenistic antecedents in stimulating the design by Giovanni Branca in 1629 of combinations of air or steam jets and rotor wheels, devices which soon found their way back to China as we shall see presently (p. 144). However, the water-mill would have been the dominant influence in his mind for whether his jet was the hot air rising through a cowl and tube from a furnace or the steam blown forth from a *sufflator* or steam-blowing head, it was always in line with the plane of rotation and not at right angles to it. The zoetrope-spit was more closely allied to the Chinese helicopter top (p. 281) and the Western vertical windmill, whence surely descended the screw propeller and the aircraft propeller. For these reasons it is not impossible that some Chinese inventor may have played a part in the introduction of screw propulsion in ships. Indeed, there is a story that a model of a Chinese screw propeller was brought to Europe and seen by a Colonel Mark Beaufoy about AD 1780, who described it a time when screw-propulsion was very much in the air, if not in the water. For our part, we have only to note that the Chinese sculling-oar (the *yuloh*, see pp. 206 ff in volume 3 of the abridgement) has affinities with the screw and that, as we have just seen, tangent-plane helicoid structures were very much at home in Chinese culture. It is not inconceivable, therefore, that some Chinese artisan thought of making a toy boat move by powering a set of helicopter vanes underneath it. Such a contribution to the main stream of descent of the screw-propeller from Archimedes through Leonardo would not be unworthy of recognition.

SPRINGS AND SPRING MECHANISMS

The elastic properties of bamboo laths have already been mentioned and it is sure that the Chinese made good use of them from an early time. Springs were certainly employed in the numerous mechanical toys and automata of which a brief account will shortly be given (p. 101). Other substances such as horn and sinews were used in the construction of bows and crossbows, while springs appear too, in varieties of pole lathe, in simple devices such as door closers, and traps for wild animals.

Though compound springs made of many leaves had been familiar from late Zhou [Chou] (eighth century BC) onwards in the form of crossbows, their application to vehicles as cart-springs never became general, though there are indications that such an invention was made in the seventh century AD (see p. 162 below). In any case Europe was much more backward in the use of springs and leaves. The crossbow was only known as a form of artillery weapon in Alexandrian times, and the use of leaved springs on vehicles did not occur until the end of the sixteenth century AD, though spring clocks and watches start from AD 1480. Of course, springs on vehicles did not become essential until, with the early locomotives, the wheels themselves had to ensure a permanent and simultaneous contact with the railway track.

Torsion springs must have been known in China from early times, since they were used in the frame-saw (p. 27 above). Metal springs occur in spring padlocks (see p. 153 below). Closely related to the spring is the vibrating wire, put to such good use in the cotton bow for loosening and separating fibres of the plant material instead of using a carding machine; but this is probably an Indian technique which came into China with cotton itself.

One of the most remarkable employments of springs in medieval China was for tripping the mechanical figures which appeared and sounded bells, gongs and drums to mark the passage of time (see below, pp. 220 ff).

CONDUITS, PIPES AND SIPHONS

In all the the previous sections in this chapter, we have been concerned with the utilisation and transmission of mechanical energy. But one of the things human beings most desire to do, from the earliest stages of technology, is to transmit liquids and gases from place to place. The whole subject of water supplies and the engineering of artificial canals dug cross-country through earth and rocks logically belongs here, but in China (as before in Egypt and Mesopotamia) it was so important that we must reserve for it a special discussion (see volume 5 of this abridgement).

Conduits or flumes of a more modest nature contrived in wood or split bamboo were always abundantly used in China for small-scale farm irrigation systems (see Fig. 291) and also for mining alluvial tin. According to the *Tang Yu Lin* [*Thang Yü Lin*] (Miscellanea of the Tang Dynasty) 'everyone says how

Fig. 291. Irrigation conduit constructed in wood. From the *Tu Shu Ji Cheng* (Imperial Encyclopaedia) of AD 1726.

good the people of Longmen [Lungmên] are in making hanging channels for water; they lead it up and down as if by magic'.

The history of piping may be followed in various writings. The important aqueducts so characteristic of civil engineering in European antiquity show that the principle of the inverted siphon was well known there, carried across the landscape on viaducts of stone or brick, but these have no close parallels in the technique of the Chinese. In the West, generally speaking, pipes of hollowed wood were used from the second millennium BC in Egypt, and copper tubing at least as early, while lead pipes, longitudinally soldered, were quite common in Roman cities. There were, however, bronze pipes in the water supply of Pergamon which involved the use of two inverted siphons; these sustained a pressure of 20 atmospheres (760 torr).

Dr Needham has not been able to find any instance of the use of metal piping in truly eotechnic China, but Nature offered there a material which was admirably adapted for the same purpose and unexpectedly strong, though perishable, namely the stems of bamboo. It may well be that the earliest large-scale use of this took place in the Sichuanese salt-fields, for brine, unlike fresh water, will not permit the growth of algae and consequent rotting of the tubes. Fig. 292 shows brine 'mains' in modern times. The joints in the tubing are

Fig. 292. Brine conduits in bamboo piping at the Ziliujing salt-fields in Sichuan, 1944. Photographed by Cecil Beaton.

sealed with a mixture of tung-oil and lime. From rubbings of Han bricks which show the salt-industry it seems certain that the bamboo pipelines were already in full use at that time. For agricultural purposes bamboo piping was used, but it needed frequent replacement. However, references to piped water-supplies for palaces, houses, farms and villages are not uncommon. The largest systems of this kind appear to be the water-mains made of large bamboo trunks installed at Hangzhou [Hangchow] in AD 1089 and at Guangzhou [Canton] in 1096. Caulked in the usual way and lacquered on the outside, holes were provided at intervals for freeing blockages, and there were ventilator taps for the removal of trapped air. These seem to have been due to the great poet-official Su Dongpo [Su Tung-Pho], who as a Sichuanese knew of the brine pipelines in his own province.

At Canton some of the sections were made of earthenware piping, and it now seems that the wide use of piped water-supplies in ancient and medieval China has been greatly underestimated. From archaeological excavations many different kinds of piping have come to light. Ancient buildings near the tumulus of the first emperor Qin Shi Huang Di [Chhin Shih Huang Ti] in the Wei valley have yielded third century BC water-conduits of thick stone-ware of pentagonal cross-section (Fig. 293 *a*). Of the same period are stone-ware well-linings about one metre in diameter and each half a metre long found at Xianyang [Hsienyang]. Again, there are fine examples of Han earthenware piping, fitting together with male and female flanges, and including sharply bending right-angle pieces (Fig. 293 *b*). These pipes continue with little change through the Han and Tang. Occasionally too pipes were made for air. A rock-cut Daoist temple above the Lindong [Lintung]

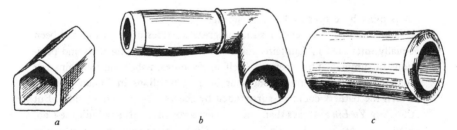

a   *b*   *c*

Fig. 293. Ancient and medieval Chinese water-pipes. (*a*) Section of thick stoneware piping from a Qin site near Lindong. Length 0.6 metres, bottom width 0.46 metres, sides about 0.3 metres, thickness *c.* 7.6 cm; (*b*) straight section and right-angle bend in stoneware from a Han site near Xianyang. Length 0.46 metres, internal diameter from 20 to 23 cm; (*c*) straight section found at Tang sites in and near Xian and Loyang, with male and female flanges for jointing. Length 0.4 to 0.46 metres, internal diameter 20 to 23 cm.

gardens is kept perpetually cool in summer by a tube bringing cold air from some mountain cleft.

Of the siphon, something has already been said when discussing water-clocks (see volume 2 of the abridgement), and here we just note that it was certainly as widely used for automata in China as in Alexandria. However, it must be remarked that the Chinese words for siphons may sometimes have referred to syringes or pumps.

Allied to siphons and pumps, there is the question of the existence of fountains in traditional China. It is now clear that evidence for these is to be found from almost every century after the Han. The technology was available and various notable examples may be quoted. Four hundred years before the building of the famous Yuan Ming Yuan palace, the last emperor of the Yuan dynasty, Toghan Timur (ruled AD 1333 to 1367) had surrounded himself with mechanical toys of all kinds, clocks with elaborate moving figures, and fountains of several different sorts. There were dragon-fountains with balls kept dancing on jets, dragons spouting perfumed mist. Yet some two hundred years before, Meng Yuanlao [Mêng Yuan-Lao], describing in AD 1148 the glories of Kaifeng [Khaifêng], the capital lost to the Jin [Chin] Tartars, tells us that at a certain temple there were:

> two statues of the Buddhas Manjuśri and Samantabhadra riding on white lions. From the five fingers of each of their outstretched hands, which quivered all the time, streams of water poured in all directions. For this purpose wheels were used to hoist the water up to the top of a high hill behind, where there was a wooden cistern. At the appointed times this was released (through pipes) so that it sprayed like a waterfall.

This must have been well worth seeing.

Four hundred years earlier still, the great worthies of the Tang had been equally interested in fountains and similar means of cooling halls and pavilions in summer. This does not itself imply upward-shooting fountains but something more like those lodges or bathing pavilions in Indian lands in which the bathers could sit surrounded by sheets of water on all sides. But the *Tang Yu Lin* also says that when the emperor Ming Huang built the Cool Hall about AD 747 Chen Zhijie [Chhen Chih-Chieh] submitted a memorial to the throne admonishing most severely against it on grounds of extravagance. Chen was summoned to court and at a time when the heat was really extreme. It then reports:

> The emperor was in the Cool Hall and behind his seat the water struck the fan-wheels while cool air played around one's neck and

clothes. Chen Zhijie arrived and was given a stone chair. A low thunder growled. The Sun was hidden from sight. Water rose in the four corners and forming screens fell again with a splash. The seats were cooled with ice, and Chen was served with marrow-chilling drinks, so that he began to shiver and his belly was filled with rumblings. Again and again he begged permission to leave, though the emperor never stopped perspiring, and at last Chen could hardly get as far as the gate before stopping to relieve nature in the most embarrassing way. Next day he recovered his equanimity. But people said that 'when one discusses affairs one should deliberate thoroughly on them first, and not put oneself in the emperor's place'.

For fountains this must suffice. Of the fan-wheels kept rotating behind the emperor's seat, we shall learn shortly.

VALVES, BELLOWS, PUMPS AND FANS

Now we must look at the complex early invention connected with propelling liquids (mainly water) and gases (mainly air) along tubes. Of pushing mechanisms there were many kinds; the flexible animal-skin bag, the piston working in a cylinder, and the rotary fan. But the common feature in all machines of this kind except the last was the presence of clack-valves, which needed to be no more than small hinged doors covering the exits and entrances of pipes in the walls of the propulsion chamber.

Universal in China for every use by artisans, even on a larger scale for minor industries, is the box bellows (Fig. 294). This surpasses in efficiency any other air-pump made before the advent of modern machinery. From the sectional drawing, it can be seen that the box-bellows is a double-acting force and suction pump; at each stroke, while expelling the air on one side of the piston, it draws in an equal amount of air on the other side. Whenever this bellows first came into general use it provided that fundamental necessity in metallurgy, a continuous blast of air. Note that the piston is packed with feathers (the ancestors of piston-rings). No less than twelve illustrations in the *Tian Gong Kai Wu* [*Thien Kung Khai Wu*] (The Exploitation of the Works of Nature) of AD 1637 shows its use by metal workers; Fig. 295 illustrates it in a bronze foundry. The common Japanese bellows, though similar, is less ingenious, since the piston carries a valve, and the blast occurs only on the push and not on the pull.

It has been much admired, being essentially equivalent to the Alexandrian double-barrel force-pump for liquids with the two cylinders elegantly combined into one. If pipes were connected to the intakes it would become the

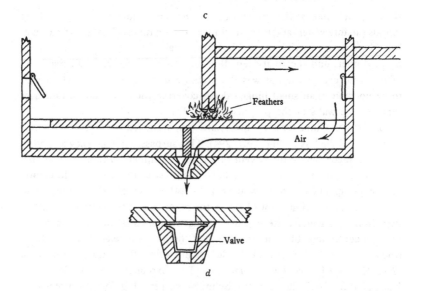

Fig. 294. The Chinese double-acting piston bellows (above). Beneath is a lengthways section to show the arrangement which ensures a continuous blast. Air enters alternately at each end of the rectangular cylinder. For a fuller description of the action, see text. The valves with their pivoting pegs resemble in form the characteristic Chinese house-doors. The swinging outlet valve is seen in the photograph in the neutral position and is at the side of the cylinder.

Fig. 295. A battery of double-acting piston bellows in use by bronze founders. According to the caption they are casting a tripod cauldron, and the channels for the molten metal are marked as made of clay. From *Tian Gong Kai Wu [Thien Kung Khai Wu]* (The Exploitation of the Works of Nature) of AD 1637. (Hereafter referred to by its translated English title.)

pump of de la Hire (AD 1716), and its connection with the later steam-engine cylinder of James Watt (where steam enters at each end so as to create a vacuum on each side of the piston alternately) is obviously one of close formal resemblance. It could also have served the same purpose as Robert Boyle's and Robert Hooke's air-pump (1659).

It is difficult, unfortunately, to bring forward much evidence as to the precise antiquity of this machine. Bellows for metal-working played a very important part in ancient Chinese thought and mythology (see p. 105 in volume 1 of this abridgement), while the text of the *Zuo Zhuan* [*Tso Chuan*] (Master Zuoqui's Tradition of the *Spring and Summer Annals*) which deals with the period 722–453 BC, uses an expression which may mean 'iron blown by the bellows'. Again, from the *Mo Zi* [*Mo Tzu*] (The Book of Master Mo) it is clear that at the beginning of the fourth century BC it was customary to use bellows to direct toxic smoke against troops attacking cities, or blown into the openings of enemy tunnels. These bellows certainly had a mechanisation of the push-and-pull motion alternating between two pots or cylinders, similar to the double-cylinder force pump of Heron of Alexandria (first century AD). From this point it was not a far cry to combining the pistons and cylinders into one, though we know nothing of the ingenious inventor. However, there is a curious hint that this may have occurred as early as the fourth century BC. The *Dao De Jing* [*Tao Tê Ching*] (Canon of the Dao and Its Virtue) (conservatively of this date) says;

> Heaven and Earth are all that lies between,
> Is like a bellows with its tuyère (blast-pipe);
> Although it is empty it does not collapse
> And the more it is worked the more it gives forth.

The statement in the third line could hardly have been made of any skin bellows, but would clearly apply to the piston variety whether the latter was hinged or straight-sliding. Commentators from the third century AD to eighteenth century all say that what Lao Zi [Lao Tzu] was referring to was the 'push-and-pull bellows'.

Evidence for the existence of the piston air-pump in the Han may perhaps be derived from the *Huai Nan Zi* [*Huai Nan Tzu*] book. Here there is a complaint about the decline of primitive simplicity, and among the extravagances of the age 'bellows are violently worked to send the blast through the tuyères in order to melt the bronze and the iron' and a commentator living about AD 200 explains that the reference is to 'push-and-pull' bellows.

Certain anthropological facts strengthen this opinion. We are familiar with the fact that when one inflates a bicycle tyre the lower end of the pump becomes hot. This fact was discovered by the primitive peoples of south-east

Asia (especially the Malayan-Indonesian region), who made use of it in one of the most remarkable of eotechnic devices, the piston fire-lighter. This is a pump at the bottom of which tinder is placed, and if indeed their stock-in-trade, then the Chinese piston-bellows may be regarded, at least as a working hypothesis, as derived from it. Indeed piston-bellows are rather widespread in the primitive cultures of East Asia.

Generally speaking, piston-pumps for liquids were not a feature of the Chinese eotechnic tradition. Yet there had been one element of traditional art which involved a principle near to that of the suction-lift pump, namely the long bamboo tube-buckets which were being sent down, from Han times onward, to the brine at the bottom of boreholes of the Sichuan salt-field (Fig. 292). These buckets carried a valve at the base by which they were filled, and would have constituted suction-lift pumps if they had fitted tightly to the walls of the borehole. But the Chinese aim was different; it must be remembered that the contents had to be raised some 300 to 600 metres, not spilled out after a short haul, and all within the limits which atmospheric pressure could fill. Observations made in the 1920s, showed that the filling time at the brine was 180 sec., the emptying time at the borehead 300 sec., the raising time 25.5 min. for each load, the dimensions of the buckets 25 metres long by 7.6 cm in diameter, and the contents 132 kg. This was a considerable engineering operation.

The relationship of the brine-bucket valves to valves in air-pumps or bellows was perfectly appreciated by Su Dongpo [Su Tung-Pho] in a passage written AD 1060. In his description of the Sichuan [Szechuan] salt industry, he says:

> They also use smaller bamboo tubes which travel up and down in the wells; these cylinders have no (fixed) bottom, and possess an orifice in the top. Pieces of leather several inches in size are attached (to the bottom, forming a valve). As these buckets go in and out of the brine, the air by pushing and sucking makes (the valve) open and close automatically. Each such cylinder brings up several *tou* of brine. All these boreholes use machinery (hoists). Where profit is to be had, no one fails to know about it.
>
> The *Hou Han Shu* speaks of 'water(-driven) bellows' This is applied to iron-working in Sichuan, and large ones are used. It seems to me to be the same kind of method as that used in the brine-collecting tube-buckets of these salt-wells. Prince Xian [Hsien] (who made a commentary on the *Hou Han Shu* in the Thang dynasty) did not understand this, and his ideas on the subject were wrong.

This is most valuable evidence on a number of points. Since Su Dongpo identifies valves in bellows working like those of the buckets in the shafts, the piston-bellows in some form or other must have been fairly familiar in his time, and though water-driven piston bellows were probably less common, he speaks as if he had seen them. He reproaches Prince Xian for wanting to substitute words meaning 'leather bellows' for the 'push-and-pull' of the text. Then, just over a century later, we find a further reference to piston-bellows in one of the works of the great Neo-Confucian philosopher Zhu Xi [Chu Hsi].

Finally one can actually illustrate piston-bellows from the century following Zhu Xi, for a book printed in AD 1280 gives two small pictures of smiths working at their anvils with unmistakable piston-bellows by their side. We can thus quite safely conclude that the piston-bellows was well known in the Song. Yet on p. 90 reasons were given for thinking that it probably goes back much further, probably long before the Tang.

Though piston-pumps for liquids were not prominent in Chinese eotechnic practice, there is sometimes reason to suspect their presence, and sometimes rather extraordinary examples come to light. Let us consider first the simplest ancestor of all such pumps, the syringe. In its most primitive form, a tube of bone or metal fixed into a bag of animal origin, it was closely similar to the primitive skin-bellows already mentioned (see above and pp. 87 and 90). Piston syringes seems to begin with the Alexandrians, for Philon alludes to the squirting of rose-water, and a very clear description of a bronze syringe occurs in Heron. Roman syringes exist in museums. From accounts of Indian surgical equipment it would seem that it developed in that civilisation at least as early. There it is also associated with a great folk-festival, Holi, when people squirt coloured water and perfumes at each other. In China there is nothing like this, but the instrument is certainly ancient there. From its modern name of 'water-gun', a late appearance might be suspected, but there were probably a number of other terms which became obsolete. The *Wu Jing Zong Yao* [*Wu Ching Tsung Yao*] (Collection of the Most Important Military Techniques) of AD 1044, preoccupied at this point by fire-fighting says:

> For syringes one uses long pieces of (hollow) bamboo, opening a hole in the bottom (septum) and wrapping silk floss round a piston-rod inside (to form the piston). Then from the hole water may be shot forth.

This is much more significant than it might seem, for the use of bamboo emphasises once again the cardinal importance of this material for all initiatives concerning tubing in classical Chinese technology.

In the eleventh century AD, however, the view is much clearer. The military encyclopaedia just mentioned gives us elsewhere a very remarkable account of a flamethrower for naphtha (Greek fire, in fact), which constituted a liquid piston-pump of ingenious design (Fig. 296). A translation of part of the passage could not be omitted here. It runs:

> On the right is the naphtha flamethrower (lit. fierce fire oil
> shooter). The tank is made of brass, and supported on four legs.
> From its upper surface arise four (vertical) tubes attached to a
> horizontal cylinder above; they are all connected with the tank.
> The head and the tail of the cylinder are large, (the middle) is of
> narrow (diameter). In the tail end there is a small opening as big as
> a millet-grain. The head end has (two) round openings 3.8 cm in
> diameter. At the side of the tank there is a hole with a (little) tube
> which is used for filling, and this is fitted with a cover. Inside the
> cylinder there is a (piston-)rod packed with silk floss, the head of
> which is wound round with hemp waste about 1.3 cm thick. Before
> and behind, the two communicating tubes are (alternately) occluded
> (lit. controlled), and (the mechanism) thus determined. The tail has
> a horizontal handle (the pump handle), in front of which there is a
> round cover. When (the handle is pushed) in (the pistons) close the
> mouths of the tubes (in turn).
>   Before use the tank is filled with rather more than three catties
> [1.8 kg] of the oil with a spoon through a filter; at the same time

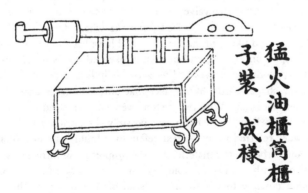

Fig. 296. Illustration of military flame-thrower for naphtha or Greek fire (distilled petroleum), eleventh century AD. This is a liquid piston-pump of ingenious design, recognisably Chinese in character. It is labelled 'Drawing of the complete Fierce Fire Oil (Projector, with its) horizontal barrel and its tank'.

gunpowder (composition) is placed in the ignition-chamber at the head. When the fire is to be started one applies a heated branding-iron (to the ignition-chamber), and the piston-rod is forced fully into the cylinder – then the man at the back is ordered to draw the piston-rod fully backwards and work it (back and forth) as vigorously as possible. Whereupon the oil (the naphtha) comes out through the ignition-chamber and is shot forth as blazing flame.

This is a text of great interest, dating as it does from a couple of decades before the time of William the Conqueror. We know indeed that the flame-thrower was already in use in the first years of the eleventh century AD from a story in the *Qing Xiang Za Ji* (*Chhing Hsiang Tsa Chi*) (Miscellaneous Records on Green Bamboo Tablets), which tells how certain officials were laughed at for being more expert with it than with their writing-brushes. The description reads in part like a set of army instructions but is none too explicit about the details of the internal mechanism – perhaps these were 'restricted'. We can, however, be confident that the purpose of the four upright tubes was to enable a continuous jet of flame to be shot forth, just as the double-acting piston-bellows gave a continuous blast of air, and the most obvious way of effecting this was to have a pair of internal nozzles one of which was fed from the rear compartment on the backstroke. The reconstruction which Dr Needham believes most probable is shown in Fig. 297 *a, b*. Such a design is very compatible with the directions in the text that the machine was to be started with the piston-rod pushed fully forward, and it also agrees with the statement that the 'two' communicating tubes (i.e. the feed-tubes) are alternatively shut off. Only two valves were necessary since the pistons them-selves acted like slide-valves on the feeds, and the whole device was more suitable for a light fluid like naphtha than it would have been for water, since the feeds were open only at the end of each stroke and response had to be rapid. That the pistons themselves had no valves is indicated partly by their long and narrow shape, partly by the fact that only one central feed-tube would have been necessary. Why two pistons were fitted instead of a single one, as in the box-bellows, is hard to say; possibly for greater rigidity.

The gunpowder referred to in the text refers to sulphur, saltpetre (potas-sium nitrate) and carbonaceous material, but the proportion of nitrate in some of the earliest eleventh century AD compositions was so small that it would have been quite possible to use them as a kind of slow-match in the way described.

If the military engineers of the Song could produce such elegant pumps to withstand the attacks of enemies such as the Jin Tartars and later the Mongols, why did piston water-pumps seem new in the seventeenth century AD? It has been suggested that this was because the square-pallet chain

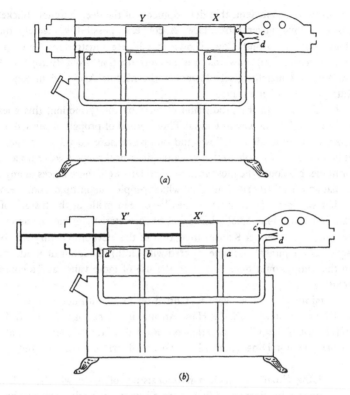

Fig. 297. Reconstruction of the mechanism of the double-acting double-piston single-cylinder force-pump seen in Fig. 296. (*a*) at the end of the forestroke; (*b*) at the end of the back-stroke. The cycle may be visualised as follows:

(1) When the piston is fully pushed in (piston positions *X* and *Y*), the naphtha from the forward compartment (or air, if starting) has been fully expelled through the nozzle *c*. Naphtha enters the rear compartment through the feed *b*, drawn up by the partial vacuum produced by the closure of valve *d'*.

(2) On the back-stroke feed *b* is shut, valve *d'* opens, and naphtha is expelled through nozzle *d*. A partial vacuum is produced in the forward compartment by the closure of valve *c'*.

(3) When the piston-rod is fully withdrawn (piston positions *X'* and *Y'*), the naphtha from the rear compartment has been fully expelled through nozzle *d*. Naphtha enters the forward compartment through feed *a*, drawn up by the partial vacuum produced by the closure of valve *c'*.

(4) On the fore-stroke the feed *a* is shut, valve *c'* opens, and naphtha is expelled through nozzle *c*. Thus the cycle is continually repeated until the tank is dry.

The ignition chamber contained a low-nitrate gunpowder composition to act as a piece of slow-match.

pumps were so efficient, that developments of the deep borehole buckets and piston-bellows never took place. A converse possibility, namely that the double-acting piston-bellows shown in Chinese illustrations from the thirteenth century AD onwards was an introduction from Europe is highly unlikely, for metallurgical blowers in the Middle Ages and in Renaissance Europe used other means.

Turning now to the production of blast by rotary motion, this goes back a surprisingly long way in China. This method of propelling air stands apart from all other kinds of bellows and pumps because no valves are necessary; it is similar in principle to the way a paddle-wheel propels water up a channel or flume. No doubt the most ancient of all fans were those pieces of any handy or flat and relatively rigid material which people caught up to cool themselves in hot summers. Interesting studies have been made of the history of hand fans in China. The radial folding fan, so characteristic of East Asian cultures, seems to have been a Korean invention of the eleventh century AD, but the action of the punkah (a frame covered with cloth), though little used, and only in the south, could have suggested the use of more vanes and a continuous rotation.

Certainly such vanes were first fitted on to a continuously revolving axle in China not later than the Han. An important passage in the *Xi Jing Za Ji* [*Hsi Ching Tsa Chi*] (Miscellaneous Records of the Western Capital) refers to the inventor Ding Huan [Ting Huan] (flourished 180 AD) and says:

> Ding Huan also made a fan consisting of seven wheels, each three metres in diameter. They were all connected with one another, and set in motion (by the power of) one man. The whole hall became so cool that people would even begin to shiver.

This must have made Han palaces, with their winding waterways, very pleasant in the heat of summer. But air-conditioning was not confined to them. From a passage quoted above (p. 86) we found it was used in Tang [Thang] palaces and from numerous mentions during the Song (eleventh and twelfth centuries AD) the refrigerant effects of artificial draught seem to have been appreciated even more widely.

But when we turn to the rotary-fan winnowing-machine new facts emerge. The problem the Chinese faced was to find a way of substituting a controllable air-current for the natural breezes on which the farmers of old had depended for separating the chaff from the grain. Metal workers had had their blast from remote antiquity, but the farmers needed a gentler one, and they solved the problem in a different way. The encyclopaedias of rustic science show two forms of the winnowing machine, one in which the crank-operated fan is set high up and quite open (Fig. 298), the other in which it is fully

Fig. 298. Open treadle-operated rotary-fan winnowing-machine. This form must be at least as ancient as the Han period.

enclosed in a cylindrical casing and made to direct its draught over a grain-chute under a hopper (see Fig. 287). Though this latter apparatus looks so like those which we are accustomed to seeing on farms of our own time, we must remember that its date is only just after 1300 AD. From the greater sophistication of the closed type, one would be inclined to regard it as the later, while the open type would seem more appropriately to belong to the the lifetime of Ding Huan. But there is some evidence for the closed type as early as the Han, while the open may be much older than he. Dr Needham has never seen the open type, nor any contemporary photograph of it, but the closed-box type is still in widespread use. Of course the ancient method of throwing up the grain in baskets still persists in some areas.

What Wang Chen had to say in AD 1313 in the *Nong Shu* about the rotary-fan winnowing-machine (see Figs. 287, 298) is worth reading.

> The rotary winnowing fan (*yang shan*, 颺扇).
>
> According to the *Ji Yun* [*Chi Yün*] (Complete Dictionary of the Sounds of Characters) *yang* means 'flying in the wind', and the *yang* fan is a machine for winnowing grain. To make it one puts at the centre (of a box) a transverse axle fitted with four or six vanes made of thin boards or of bamboo (slips) glued together. There are two types, one with the fan vertically mounted, the other having it horizontally mounted, but both include a driving shaft worked either manually or by means of a treadle, in accordance with which the fan rotates. The mixed grain and chaff from the mortar or the roller-mill is put into the hopper, communicating at the bottom with a separator, through which the grain falls down as fine as the holes in a sieve. As the fan turns, it blows away the husks and bran; thus the pure grain is obtained.
>
> Some people raise the fan high up (without enclosing it) and so winnow; ... these machines are much more efficient than the throwing up in baskets.
>
> As Mei Shengyu [Mei Shêng-Yü] the poet says:
>
> 'There on the threshing-floor stands the wind-maker,
> Not like the feeble round fans of the dog-days,
> But wood-walled and fan-cranked, a cunning contrivance,
> He blows in his tempest all the coarse chaff away,
> Easy the work for those manning the handles –
> No call to wait for the weather, the breezes
> To free the fine grain from its husks, that our fathers
> Needed for tossing their baskets on high.'

Thus at once we find evidence for the existence of the machine early in the eleventh century AD, for Mei Shengyu died AD 1060. But pressing further backwards, we can establish it in the early part of seventh also, for that was the time when Yen Shigu [Yen Shih-Ku] was writing his commentary on the *Ji Jiu Pian* [*Chi Chiu Phien*] (Dictionary for Urgent Use), which the Han scholar Shi You [Shih Yu] had put together about 40 BC. Discussing it, Yen Shigu [Yen Shih-Ku] adds some synonyms for rotary grain-mills and goes on to say:

The fan (*shan*, 扇) means the rotary winnowing-fan (*shan che* [*shan chhê*], 颺扇), and *tui* [*thui*] (隤) explains the principle of the *shan che* [*shan chhê*]. Some write this *tui* [*thui*] (遀), but in any case it means 'to fall', that is to say, while (the machine) fans, (the grain) falls through. Other people after pounding toss up in winnowing-baskets, to blow away the chaff.

These words might thus seem to place the winnowing-machine not only in the early Tang, but also in the latter part of the Early Han (first century BC). Yet how valid was Yen Shigu's interpretation of Shi You? Strong support for it can be found in those Han tomb-models already mentioned and an example illustrated in Fig. 288, which show what looks extraordinarily like a winnowing-machine with its hopper and crank-handle. There is also the evidence that Ding Huan was using the principle for other purposes in the second century AD. But certain Han bricks show another device. One man stands behind a pair of uprights some 1.5 metres high having long flat fan-boards on them which he seems to be working quickly to and fro, while another man in front of them shakes out grain and chaff from a basket held high above his head. We need to know more about this Han winnowing method, and whether it was oscillatory or rotary, but in any case some later forms of it may well have been the background for the remark by Wang Zhen [Wang Chen] about 'horizontally mounted fans', of which unfortunately no illustration has come down to us. All in all, however, we need have no hesitation in placing the principle of rotary blowers in the Han time, and perhaps the very early Han (second century BC).

This contrasts in a remarkable way with the situation in Europe. If it is correct to say that the earliest rotary blowers in Europe are those pictured by Agricola for mine ventilation in the mid sixteenth century AD, then it is very hard to believe that the idea did not travel west from China. A striking feature of the Chinese rotary fans is that the air-intake is always shown as central, so that they must be considered the ancestors of all centrifugal compressors. Even the great wind-tunnels of today derive from them.

As for the enclosed rotary-fan winnowing machine, Europe acquired it even later. Special investigations have shown that it was introduced to the West from China early in the eighteenth century AD as part of a distinct series of waves of agricultural transmissions in alternate directions.

To sum up, historically the precursor of the Chinese single-cylinder double-acting piston air-pump was the bellows of skin employed by the metallurgical artisans of the Zhou [Chou] period (eleventh to third centuries BC). At this same time the peoples of south-east Asia knew the piston firelighter, and probably the air pump was born from their union, though the origin of the valves remains obscure.

In the Warring States (fourth century BC) oscillating levers were used to work pairs of bellows or pumps, and Han writings from the second century BC afford references which point to some kind of push-and-pull bellows which were non-collapsible and therefore probably of piston type. By the Song (eleventh century AD) the characteristic air-pump is already in its mature form, implying a prior presence in the Tang (seventh to tenth centuries). So much for the propulsion of gaseous matter by pistons.

Where liquids were concerned the valved buckets used already in the Han for raising brine from deep boreholes approximated to the suction lift pump. There are also strong indications that pumps of this kind, masked now by confusing terms, were constructed in the Later Han (second century AD). By the Song again we have a remarkable use of the piston-pump in the military flame-thrower. Yet owing perhaps to the universal use of the simpler chain-pumps, the piston-pump for water was uncommon or absent in the Chinese Middle Ages.

Finally, rotary blowers make their appearance early, especially in the practical form of the rotary fan winnowing machine, another typical piece of Chinese technology. It seems certain that all European rotary gas-blowers derive from this, but to the Renaissance West must be attributed the extension of the principle to the propulsion of liquids.

3

Mechanical toys and machines described in Chinese works

If there was one field more than any other in which all the basic mechanical principles discussed in the previous chapter were used together, it was that of providing mechanical toys, puppet plays, trick vessels and so on, for the amusement and prestige of successive imperial courts. In our brief survey, besides coming across the names of mechanicians known only for this work, we shall also meet many of the best-known engineers as well. Moreover, the subject is one connected with two bodies of semi-legend, automata on the one hand, and flying machines on the other. We have already had something to say of the former (volume 1, p. 92 of the abridgement) and shall treat lightly of the latter below (p. 275).

The wealth of mechanical toys in treatises of Alexandrian times, especially of Heron, is well known. There are figures carrying out all kinds of motions, birds singing and temple doors opening and closing, while there is also a text about a puppet theatre. In later centuries the construction of such ingenious contrivances became a speciality of the Indians and the Arabs, who were particularly interested in automata fitted to striking water-clocks and in mechanical cup-bearers. Here they may have had some inspiration from China, for as we shall see, these were notable in Sui and Tang times. The mechanical doves and angels of Villard de Honnecourt in Europe of the thirteenth century AD were in the same tradition, which continued into the eighteenth century, till in our own time we find ourselves surrounded at every stage of life with a thousand gadgets from model railways and model aircraft to computer games of 'virtual reality', which would in earlier ages have been the marvellous secrets of imperial courts.

It seems doubtful whether the Chinese mechanical toys were ever inferior to those constructed by the Alexandrians and the Arabs. The theatrical connection is also equally clear in China. Some believe that the idea of animating such puppets originated from the thought of bringing to life the wooden or clay models of human beings which the Han peoples placed in their graves as servitors of the dead. It is also urged that many of the enigmatic scenes and designs found on Han tomb-carvings and mirrors represent mechanical toys of various kinds. Indeed, there is an early story of 206 BC, but available to us only in a source of the sixth century AD, the *Xi Jing Za Ji* [*Hsi China Tsa Chi*] (Miscellaneous Records of the Western Capital), which deals with a mechanical orchestra of puppets which the first Han emperor found in the treasury of Qin Shihuangdi [Chhin Shih Huang Ti].

> There were also twelve men cast in bronze, each 1 metre high, sitting upon a mat. Each one held either a lute, a guitar, a *sheng* or a *yu* (mouth-organs with free reeds). All were dressed in flowered silks and looked like real men. Under the mat there were two bronze tubes, the upper opening of which was a metre and more high and protruded behind the mat. One tube was empty and in the other there was a rope as thick as a finger. If someone blew into the empty tube, and a second person (pulled upon) the rope (by means of its) knot, then all the group made music just like real musicians.

No air-pump or bellows seems to have been involved here. It took one person to provide the air-blast by blowing, while another set all the puppets in motion by means of cams, levers, weights, etc., all working off a central drum.

To indigenous Chinese practices were perhaps added techniques from abroad; there were interchanges of conjurers and acrobats between Han China and Roman Syria, and some Hellenistic mechanical items may have accompanied them.

One of the most circumstantial accounts of puppets which has come down to us from Han times relates to the work of the famous engineer Ma Jun [Ma Chün] who flourished in the San Guo period (227 to 239 AD). In the *San Guo Zhi* [*San Kuo Chih*] (History of the Three Kingdoms) we read:

> Certain persons offered to the emperor a theatre of puppets, which could be set up in various scenes, but all motionless. The emperor asked whether they could be made to move, and Ma Jun said that they could. The emperor asked whether it would be possible to make the whole thing more ingenious, and again Ma Jun said yes, and accepted the command to do it. He took a large piece of wood and fashioned it into the shape of a wheel which rotated in a horizontal position by the power of unseen water. He furthermore arranged images of singing-girls which played music and danced, and when (a particular) puppet came upon the scene, other wooden men beat drums and blew upon flutes. Ma Jun also made a

mountain with wooden images dancing on balls, throwing swords about, hanging upside down on rope ladders, and generally behaving in an assured and easy manner.

But Ma Jun was by no means the only mechanician who achieved such successes in his period. Qu Zhi [Chhü Chih] was famous in the Jin for his wooden dolls' house, with images which opened doors and bowed, for his 'rats' market, which had figures which automatically closed doors when the rats wanted to leave. Ge You [Ko Yu] of Daoist sympathies, was alleged to have made an artificial sheep on which he rode away to the mountains, which probably means that he made some ingenious thing.

One would of course expect to find not only mechanical animals like this, but also actual chariots which moved of themselves. What is perhaps surprising is to find self-moving carriages attributed to the Chinese by a serious Muslim writer as late as about AD 1115. Among the commercial population in China, says al-Marwazī,

> there are men who go about the city selling goods, fruits and so on, and each of them builds himself a cart in which he sits and in which he puts . . . whatever he requires in his trade. These carts go by themselves, without any animals (to draw them), and each man sits in his cart, stopping it and setting it in motion just as he desires.

Al-Marwazī did not himself see this; he received a report of it from a traveller whose visit to China took place between AD 907 and 923. Perhaps the story is an echo not so much of Mohist legends as of what some bona fide traveller said about Chinese wheelbarrows, still in those times a quite unknown invention in the West (see p. 165 below). Or could they really have been pedal-driven carts?

In the fourth century AD Wang Jia [Wang Chia] refers to a mechanical man of jade which could turn and move, apparently of itself. But we get a clearer idea of what people were doing from the account of the masterpiece designed by Xie Fei [Hsieh Fei] and Wei Mengbian [Wei Mêng-Pien] in the *Ye Zhong Ji* [*Yeh Chung Chi*] (Record of Affairs at the Capital of the Later Zhao [Chao] Dynasty). Both worked at the court of the Hunnish emperor Shi Hu [Shih Hu] between AD 335 and 345. The text reports:

> Xie Fei also invented a four-wheeled Sandalwood Car six metres long and more than three metres wide. It carried a Buddhist statue, over which nine dragons spouted water. A large wooden figure of a Daoist was made with its hand continually rubbing the front of the

Buddha. There were also more than ten wooden Daoists each more than half a metre high, all dressed in monastic robes, continually moving round the Buddha. At one point in their circuit each automatically bowed and saluted, at another each threw incense into a censer . . . When the carriage moved onwards, the wooden men also moved and the dragons spouted their water; when the carriage stopped, all the movements stopped.

We shall meet these two inventors again in connection with more important vehicles.

Automaton cup-bearers and wine-pourers begin to be prominent in the Sui period (early seventh century AD), under the name 'hydraulic elegances'. The mechanician mostly responsible for this development was Huang Gun [Huang Kun], a man in the service of Sui Yangdi [Sui Yang Ti], at the request of whom he wrote a manual, *Shui Shi Tu Jing* [*Shui Shih Thu Ching*] (Illustrated Manual of Hydraulic Elegances) on the subject. This was edited and enlarged by his friend Du Bao [Tu Pao]. According to the accounts, these displays involved numbers of boats (about three metres long and two metres wide), fitted with mechanical devices and having moving figures on board, floating along winding stone channels and canals contrived in palace courtyards and gardens, and so passing guests in turn. All the usual beings were represented, animals and men, immortals and singing girls, playing all kinds of musical instruments, dancing and tumbling, just as in Ma Jun's time.

When we consider the possible mechanisms, we are inclined to think that the simplest way this could have been accomplished would have been to connect the boats and haul them by an endless rope or chain under the water, with the figures operated by power from small hidden paddle-wheels. Indeed, evidence for the use of such devices at this time (AD 606 to 616) is available (see volume 5 of this abridgement).

Perhaps all this grew out of a ceremony of exorcism held in Jin times (third to fifth centuries AD) on the third day of the third month, in which cups of wine were floated along little winding channels. But there was also a festival on the fifteenth day of the seventh month, when candles and lights were set afloat. Apparently from the Sui time onwards the channels were constructed indoors and fed with water from clear springs – it must have been a pretty sight to see the mechanical acrobats and cup-bearers floating serenely on their circuitous paths to the accompaniment of musical-box sounds.

In the course of time this interest in boats with mechanical figures spread to the mass of the people, or at least the more affluent of them, leading to a regular trade in model ships. In his *Peng Chuang Lei Ji Yi* [*Phêng Chhuang Lei Chin I*] (Classified Records of the Weed-Grown Window) of AD 1527, Huang Wei informs us that at Nanjing model sailing-boats were beautifully carved with crew and passengers all moving 'by means of a mechanism'. When placed in the water they would sail before the wind, and 'people who liked to busy themselves with miscellaneous affairs' engaged in competitions with them, doubtless on the lovely Hou Hu lake which reflects both the Purple Mountain and the battlements of the far-stretching city walls. By good fortune representations of model boats are preserved in extant Chinese paintings.

The Sui emperors left a general reputation for active interest in mechanical devices. There remains, for example, an account of automatically opening library doors constructed for Sui Yangdi [Sui Yang Ti] (AD 605 to 616).

> In front of the Kuan Wen Hall there was the Library, in which
> there were fourteen studies, each having windows, doors, couches,
> cushions and bookcases, all arranged and ornamented with
> exceeding great elegance. At every third study there was an open
> square door (in front of which) silk curtains were suspended,
> having above two (figures of) flying *xian* [*hsien*]. Outside these
> doors a kind of trigger-mechanism was (contrived) in the ground.
> When the emperor moved towards the Library he was preceded by
> certain serving maids holding perfume-burners, and when they
> stepped upon the trigger-mechanism, then the flying *xian* came
> down and gathered in the curtains and flew up again, while at the
> same time the door-halves swung backwards and all the doors of
> the book-cases opened automatically. And when the emperor went
> out, everything again closed and returned to its original state.

Here we have what might be considered the half-way house between the spontaneously opening temple doors of the Alexandrians and the electronically controlled doors of modern airports and supermarkets.

People in the Tang continued to feel the fascination of mechanical toys and puppet plays, and some of the latter were very elaborate, such as that which was constructed for the funeral games of a provincial governor in AD 770. Names of individual mechanicians have come down to us from this time. There was Yang Wulian [Yang Wu-Lien], afterwards a general, who made a figure of a monk which stretched out its hands for contributions, saying 'Alms! Alms!', and deposited the contributions in its satchel when they reached a certain weight. This had great success on market-days, we are told. Then there was Wang Ju [Wang Chü] who made a wooden otter which could catch fish (probably some kind of spring trap embodying a figure), and Yin Wenliang [Yin Wên-Liang] celebrated for his wooden cup-bearers and sing-song-girls who played the flute.

In the Song, glass was added to the stock-in-trade of the artisans who employed themselves in such domains, for we read of a mountain of glass with moving figures, and screens of glass behind which movements went on by the use of water-power. By this time also, puppets were involved in the service of horology. This we shall discuss later (pp. 220 ff), and here note only that this interest was fully maintained by some of the Yuan emperors.

At this point we may suitably leave the parallel European and Chinese traditions of mechanical toys. The latter may or may not have been a little younger in its origin but there never was much to choose between them for ingenuity, and when they came together in the thirteenth century AD, the European tradition did not show up to much advantage. The triumphs of the European 'Gadget Age' were yet to come.

TYPES OF MACHINES DESCRIBED IN CHINESE WORKS

We must now briefly consider the principal types of useful machines described and illustrated in Chinese books. The literature is distinctly small, perhaps mainly because the constructions of artisans, however ingenious, were too often regarded as unworthy of the attention of the Confucian literati. Nevertheless, considerable numbers of illustrations have survived from the eleventh century AD onwards (the springtime of printing), and there existed what may be called specific traditions of engineering drawing, though the draughtsmen or artists evidently did not always clearly understand what it was they were illustrating, and may have thought it beneath their dignity to enquire too closely. On the other hand, a great number of texts, many in dynastic histories, have also survived.

The difficulty with the pictures (when the veil of scholarly fastidiousness is not impenetrable) is that we know where we are technologically, but we cannot always easily get back to the original date of the illustration. Extensive research in Song editions by persons of engineering as well as sinological competence are needed, and even so the limits of what has been preserved may be reached fairly soon. That late pictures may, however, perpetuate correct technological traditions is strikingly shown by the case of a seventeenth-century AD book about silk, certain illustrations in which portray detail by detail a description of an eleventh-century text. Conversely, the difficulty with literary sources alone is that we have a firm date, which may indeed be quite early, but we are not always sure where we are technologically, either because the description of the mechanism is insufficient, or because there is ground for fearing that the meanings of the technical terms have suffered changes from time to time. The only cure for this is presumably the discovery and analysis of further texts.

Chinese technical literature has sometimes been reproached for a certain vagueness or ambiguity. What weight there is in this derives from the fact that Confucian scholars sometimes had to write about things which did not really interest them, while the technicians themselves, who could really have explained matters, did not write at all.

The nature of Chinese engineering literature

For convenience, we shall begin by describing the principal books of the agricultural family which contain illustrations. The tradition of these pictures began with admonitory pictures upon the walls of imperial palaces. The emperor Mingdi [Ming Ti] (who reigned from AD 323 to 325), himself a famous painter, left a series of pictures known as the *Bin Shi Qi Yue Tu* [*Pin Shih Chhi Yüeh Thu*] (Illustrations for the 'Seventh Month' Ode in the 'Customs of Bin' section of the *Shi Jing* [*Shih Ching*] (Book of Odes)). There is also evidence, more doubtful, that the emperor Shi Zong [Shih Tsung] (who reigned from AD 954 to 959) built a pavilion which was ornamented with scenes of tilling and weaving. In any case some such tradition certainly existed, and it seems to have had a magical significance, for the Song encyclopaedist Wang Yinglin reports a tradition that in the Tang and Song on two occasions when the palace frescoes of labour scenes were replaced by mere landscapes, there was trouble among the people and rebellions arose. In due course paintings of this kind were collected into book form, a type of publication the earliest of which was the *Yue Ling Tu* [*Yueh Ling Thu*] (Illustrations for the Monthly Ordinances (of the Zhou dynasty)). Nothing is known about it or its author, Wang Yai, but its importance is that it was the predecessor of one of the most famous books in all Chinese literature, the *Geng Zhi Tu* [*Kêng Chih Thu*] (Pictures of Tilling and Weaving).

Such was the artistic and literary importance of this work, as well as its technological interest, that its complex history and bibliography have given rise to studies in Western languages as well as Chinese. The original pictures, each accompanied by a poem, were produced by Lou Shou, an official of the Southern Song, for presentation to the emperor Gao Zong [Kao Tsung], in approximately AD 1145. Later, after Lou Shou's death, they were inscribed on stone, and probably also printed, about 1210, by his nephew Lou Yo and grandson Lou Hong [Lou Hung]. Their value to us lies in the fact that they are the oldest pictures we have of Chinese agricultural, mechanical and textile technology – with the exception of what may be gleaned from carvings in Han tombs and mural paintings of the Wei and Tang. Only military illustrations start earlier (AD 1044).

This raises, of course, the important question of the extent to which the pictures which we have today are faithful copies of the twelfth- and thirteenth-century AD originals. In AD 1696 the whole set was redrawn at imperial command by an eminent artist, Qiao Bingzhen [Chiao Ping-Chên], who followed the rules of perspective which the Jesuits had introduced from the West. The Kangxi [Khang-Hsi] emperor added a new set of poems while retaining the old ones of Lou Shou. So highly was the work prized for its symbolic significance in depicting the foundations of Chinese agrarian culture, that in 1739 a new copy of the pictures was made and new poems were written by imperial decree, to which a prose explanation was added. Then in 1742 the whole was incorporated into the *Shou Shi Tong Kao* [*Shou Shih Thung Khao*] (Complete Investigation of the Works and Days) with some new poems together with those of Lou Shou, removed in 1739, now being replaced. Fortunately, certain sets of pre-Qing illustrations have been recovered during the twentieth century. Discussion is likely to continue on the finer points of these and the many later editions, but there is no great difference between them on the essential technical matters which they represent, and we should therefore feel fairly certain that the older sets are valid for the time of Lou Shou himself.

These bibliographical technicalities, seemingly tedious, have so much importance for the comparative history of technology at the two ends of the Old World that it is necessary to dwell on their meaning for a moment. The paintings by Cheng Qi [Chhêng Chhi] must have been made close to AD 1257. The first half of the set was rediscovered and presented to the emperor soon after 1739 while the second half reached the imperial palace in 1769. The Qianlong emperor' realising their importance, had the whole engraved on stone. Whether or not the stones still exist matters the less since rubbings are available of the 'Semallé Scroll' of the *Geng Zhi Tu* (Pictures of Tilling and Weaving) of AD 1237 with all their fine detail.

If then we possess an authentic record of the Southern Song, an important contrast with Europe follows, for while the body of manuscripts of Italian and German engineers informs us about developments in the fourteenth and fifteenth centuries AD, we are here in the presence of reliable Chinese material concerning the thirteenth and fourteenth centuries. Cheng Qi must certainly have worked from the block-printed edition of AD 1237, based on original material of 1145, and during the interval Dr Needham thinks it unlikely that any technical details changed. The *Geng Zhi Tu* is usually dated AD 1237 or 1210, but its probable validity for a century earlier must be remembered. An interesting point emerges here, for all through this period (indeed from the ninth century AD onwards) China had printing while Europe did not. In other words, we can never assume the existence in the West of an invention (e.g. the crankshaft) earlier than its first manuscript evidence; but where the Chinese pictorial data are known to be based on earlier printed editions, we may feel able to give greater probability to the previous existence of the technical detail in question, so far as the limits of the bibliographical facts allow. Descriptive textual evidence, with all its peculiar difficulties of interpretation, is of course a different matter.

Whether the illustrations in the accessible editions of the *Nong Shu* [*Nung*

Shu] (Treatise on Agriculture) produced by Wang Zhen [Wang Chên] in AD 1313 are equally contemporary with his text is a matter even more important and more difficult to determine. While much attention has been given to the *Geng Zhi Tu*, only a minority of its pictures have information to give us about machines; the *Nona Shu*, on the other hand, shows us no less than 265 diagrams and illustrations of agricultural implements and machines. But here again the authenticity of the illustrative material is strongly vouched for by the significant fact that in no case has Dr Needham ever found a discrepancy between the text and the pictures. Moreover, the somewhat archaic character of the drawings, quite similar in style to the old *Geng Zhi Tu*, indicates that we may securely take them as valid for Wang Zhen's own time.

Throughout the Song, Yuan and Ming dynasties another class of literature became important, the encyclopaedias for daily use. So widespread were these that people generally considered them not worth preserving, and they are now to be found only as rare or unique copies in libraries. More than twenty of these mines of information were printed at dates ranging from about AD 1350 to 1630, and besides instructions on family customs, popular medicine and hygiene, fortune-telling and the drafting of legal documents, they also give numerous details of farming, silk technology and the arts and crafts. Thus a glance at the *Bian Yong Xue Hai Qun Yu* [*Pien Yung Hsüeh Chhün Yü*] (Seas of Knowledge and Mines of Jade; Encyclopaedia for Convenient Use), printed in 1607, finds wood-block cuts of the square-pallet chain-pump, the rotary-fan winnowing machine, the connecting-rod hand-mill, a very simple silk-reeling machine, and some looms. This 'inquire-within-upon-everything' literature invites study by historians of technology, and if the Ming editions are not likely to tell us much that we do not already know, those of the Yuan and the Song may well provide us with important new evidence.

Hence we come again upon the effects of the art of printing. Before the popularisation of typography technical monographs or treatises on, say, textiles or iron-working could have had only a very restricted circulation in manuscript, and much must have been lost, but as soon as the democratic medium got into its stride, from the tenth century AD onwards, technical subjects began to take their place, even if it was only in popular encyclopaedias, side by side with the topics which had always interested the literati. There is thus unfortunately a built-in obstacle to our quest for technological data of the Tang and earlier, and this is why the evidence of the Dunhuang fresco-paintings (seventh century AD) and similar sources has such particular importance.

During the three centuries after Wang Zhen no further contributions of importance to the literature of rural engineering appeared, but in AD 1609 a good many pictures from the *Nong Shu* appeared in an encyclopaedia, the *San Cai Tu Hui* [*San Tshai Thu Hui*] (Universal Encyclopaedia) though in a somewhat degenerate form. In spite of being during the time of the Jesuits, this was not due to any Western influence. Shortly afterwards, however, three books were prepared under their inspiration, and these broke away from the agricultural domination under which Chinese eotechnic mechanical engineering had grown up. The first was *Tai Xi Shui Fa* [*Thai Hsi Shui Fa*] (Hydraulic Machinery of the West) of 1612, published by Sabatino de Ursis (Xiong Sanba [Hsiung San-Pa]) and Xu Guangqi [Hsü Kuang-Chhi]. Then in 1627 came the *Qi Qi Tu Shuo* [*Chhi Chhi Thu Shuo*] (Diagrams and Explanations of Wonderful Machines (of the Far West)), by Johann Schreck (Deng Yuhan [Têng Yü-Han] and Wang Zheng [Wang Chêng]; this deals with a great variety of machines, including cranes, mills and sawmills, as well as water-raising engines. A shorter companion work, the *Zhu Qi Tu Shuo* [*Chu Chhi Thu Shuo*] (Diagrams and Explanations of a Number of Machines), by Wang Zheng alone appeared the same year. Owing perhaps to the abundance of illustrations in these works, they have attracted great attention.

Between 1625 and 1628, the scholar Xu Guangqi [Hsü Kuang-Chhi] occupied himself with a new agricultural compendium destined to supersede all earlier works, but it was not published until after his death, when it was edited by Chen Zilong [Chhen Tzu-Lung] as the *Nong Zheng Quan Shu* [*Nung Chêng Chhüan Shu*] (Complete Treatise on Agriculture) in 1639. This famous book is well illustrated, especially in the irrigation section, and reproduces nearly all the agricultural machinery of the *Nong Shu* with minor variations. The *Tai Xi Shui Fa* is reproduced complete, but there is no essential advance along the lines of traditional Chinese engineering beyond Lou Shou and Wang Zhen.

Contemporary with the work of Xu Guangqi was that of Song Yingxing, who produced his *Tian Gong Kai Wu* [*Thien Kung Khai Wu*] (The Exploitation of the Works of Nature), China's greatest technological classic, in 1637. Though dealing with agriculture and industry rather than engineering in the strict sense, its contents range from agriculture, milling processes, irrigation, hydraulic engineering and the silk industry, to salt and sugar technology; from bronze, iron, silver, lead, copper and tin metallurgy, to coal, vitriol, sulphur, arsenic, mercury and oil technology. They cover also the manufacture of fermented beverages, ink, paper, and ceramics as well as pearls and jade. And as if this were not enough, the book even includes military technology and transportation by carts and ships. The illustrations are the finest of any produced in China on these subjects, and in many cases the only ones, but the text does not always equal them in clarity. Early in the Qing period the work almost disappeared in China, perhaps because coinage, salt-making and weapon manufacture were government monopolies, but fortunately a copy of the original edition was preserved in Japan.

Just under a century later, the *Tu Shu Ji Cheng* [*Thu Shu Chi Chhêng*]

(Imperial Encyclopaedia) of 1726 continued to reproduce the traditional illustrations. Finally, in 1742, came the *Shou Shi Tong Kao [Shou Shih Thung Khao]* (Complete Investigation of the Works and Days), prepared by imperial order to excel all previous compendia on agriculture and agricultural engineering. It differed from its predecessors in starting with some geography and meteorology, including large sections on agricultural botany, but the engineering illustrations deviate in no way from those of the *Nong Zheng Quan Shu [Nung Chêng Chhüan Shu]* (Complete Treatise on Agriculture) of 1639.

It may be said, therefore, that by the end of the Song the Chinese tradition of agricultural engineering had reached its fullest development. The *Nong Shu* and the *Tian Gona Kai Wu* are the best sources for it, and though pumps and various kinds of gearing were introduced by the Jesuits there is no evidence that these innovations were adopted, presumably because social and economic conditions made unnecessary any changes from classical methods. Hence also there were no essential advances in the books themselves down to the beginning of the nineteenth century AD. Indeed, those who have lived in China, know that the early medieval techniques (such as the square-paddle chain-pump with radial treadle) are still in full vigour. Superseding them is bound up with problems of industrialisation, persistence of wet rice cultivation, and the like.

EOTECHNIC MACHINERY POWERED BY MAN AND ANIMALS

Here we must start with pounding and grinding procedures, probably among the most ancient of mankind's food-preparing activities. The motivating force was the need to remove the husk from cereal grains. The grain of cultivated grasses contains, besides the embryo of the future plant, a mass of starch-containing material – the endosperm – which acts as the fostering yolk until starch can be made by the new plant by the process of photosynthesis. Through the ages, this endosperm has been mankind's 'staff of life'. But it is guarded externally by the husk and granules of protein – the aleurone layer – which, when ground or pulverised, we call chaff and bran. These last also contain carbohydrate but in an insoluble and indigestible form. Many animals can digest this 'roughage' as well as the inner part of the grain. Human beings, however, deriving but partial benefit from the bran, can digest only the inner endosperm mentioned above, and even so a 'pre-digestion' in the form of cooking is necessary. The simplest and oldest form was toasting the whole grain which made de-husking possible. The grain was heated with water to about 60°C when the starch swelled and formed a kind of porridge, but the product does not keep.

If however the flour is made into a dough and baked at some 235°C the product is consistent and durable, but soon becomes stale and is difficult to eat even when thin. This led to the invention of 'leavening' by means of gas-producing yeasts. But bread will not rise properly unless advantage can be taken of the gluten proteins in the endosperm. Yet if the grain is roasted before hulling, these proteins are prematurely denatured and nothing more than porridge will result. Of all cereals, only wheat and rye can have their husks removed by threshing, and so are suitable for making aerated bread, wheat being the superior.

These facts explain the long persistence of pounding processes, variations of the pestle and mortar stemming from the mealing-stone of Paleolithic times, both in East and West. Rye was a German-Scandinavian grain which only came south a little way; bread wheat was already supreme in Greek lands by the fourth century BC, supplanting barley. Naked wheats ousted emmer, the principal husked wheat, in Italy by the second century BC, with the result that the Etruscans and Romans of the early Republic were eaters of porridge. But the Hellenistic world lived, like modern Europe, on baked bread.

In China the place which emmer and barley had in Mediterranean civilisation was taken by millet, both kinds of which, the glutinous and non-glutinous, were indigenous there. It formed the chief cereal of the Yangshao and other Neolithic cultures. But archaeological evidence shows that rice had already penetrated these cultures before the end of the third millennium BC, certainly coming from India. Actual remains in Han tombs have established the presence of wheat, barley and Job's tears (an annual grass [*Coix lacrima-jobi*] with edible seed kernels enclosed in hard shiny tear-shaped beads and indigenous to tropical Asia). But certain other cereals, notably buckwheat and the tropical grass kaoliang, are indigenous to China like millet, and were probably cultivated as early as the Zhou. Exactly when cultivation of wheat began on a large scale it is hard to say, but it was certainly a crop of the Shang period (*c.* 1400 BC), having reached China by way of Western and Central Asia, and in subsequent ages it became as characteristic of the Yellow River basin as rice was of that of the Yangzi [Yangtze]. Nor do we know when emmer was superseded by the naked wheats and though throughout later times risen bread was steamed, not baked, in China, one cannot infer from the 'steamers' present from the Neolithic onwards that leavened bread rather than other food was actually steamed in them. Probably leavened bread in China came during the Zhou, perhaps in association with many revolutionary agricultural changes of the early Warring States period. Abundance of bread flour does not depend upon rotary milling, but that invention seems to follow it fairly soon. This brings us back to engineering problems.

Pounding, grinding and milling

The simplest form of such processes, pounding (impact crushing) by means of the hand-used pestle and mortar, goes back in principle to Mesolithic times (the transitional period of the Stone Age between Paleolithic and Neolithic), or even earlier. It has continued until the present time from the Hebrides to Bali for the daily husk removal from cereal grains, as well as other purposes. It is fairly frequently represented in ancient Egyptian pictures and Greek vase-paintings. In China the pestle and mortar was used from antiquity for hulling all grains, but especially for the removal of the husks from rice grains, which are also roughly polished in the process so that they become white and shining by the removal of an outer fatty (aleurone) layer. From Han texts we know of mortars made of clay, of wood, and of stone, to which stoneware was later added. In the Tang southerners used boat-shaped pestles stamping communally in long troughs. Farmers used the pestle for breaking up clods in the fields, as apothecaries and alchemists also for their own purposes, and sometimes the pestle was (and is) suspended from a bamboo bow-spring. In a Chinese cereal-pounding scene of AD 1210 pestles are almost indistinguishable from mallets (Fig. 299).

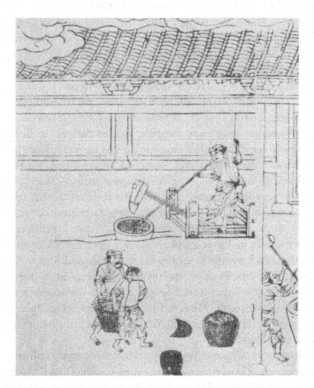

Fig. 299. The oldest extant drawing of the tilt-hammer, from the *Geng Zhi Tu* (Pictures of Tilling and Weaving) of AD 1210. From an edition of 1462.

This drawing is noteworthy, however, for the presence of something far more characteristically Chinese, namely the treadle-operated tilt-hammer. This was an extremely simple device, using lever and fulcrum to enable the pounding work to be done with the feet and the whole weight of the body, instead of with the hands and arms alone. Illustrated in all the agricultural encyclopaedias, it is one of the commonest objects of the Chinese countryside. It was used in the same way as the pestle and mortar for hulling and polishing cereal grains, but also extensively by miners in ore-dressing. In modern times it finds many applications, as in mobile earth tampers for construction works. As to its antiquity, most Chinese historians would place it without hesitation in the late Zhou, or about Qin time, though literary references to it are hard to find before the first century BC. At that time we have the *Ji Jiu Pian* [*Chi Chiu Phien*] dictionary of 40 BC and the definitions in the *Fang Yan* (Dictionary of Local Expressions) of about 15 BC attributable to Yang Xiong [Yang Hsiung]; but the best statement is that of Huan Tan [Huan Than] in the *Xin Lun* [*Hsin Lun*] (New Discussions) of about AD 20.

The pedal tilt-hammer seems not to have been used in other civilisations until much later, if at all. It is not mentioned or illustrated in Europe until an AD 1537 edition of the works of the Greek Hesiod, so it may safely be regarded as derivative there. But it seems to have had European descendants, notably a sprung treadle-operated tilt-hammer used in forges. If this goes back to the fourteenth century AD, as has been supposed, the transmission would have been as medieval as that of the blast-furnace and gunpowder.

Grinding procedures were more complex, and led further. In Europe the oldest instrument combining pressure-crushing and shearing, from the Neolithic onwards, was the grain rubber – simply a saucer-shaped stone, with a squat and bun-shaped slider movable on its upper surface. Many examples of this are to be seen in museums. Insensibly the the grain rubber developed into the saddle-quern, where a bolster-shaped upper stone rubs or rolls backwards and forwards over a large longitudinally concave lower stone. This again is often shown in Egyptian pictures, and tomb-models or toys from Egypt and Greece; it may still be seen in use in Mexico and Africa. The difficulty of feeding in the grain led to the more convenient hopper-rubber (Fig. 300) in which the upper stone has carved within it a hopper with a slit for the grain to pass through. From this there developed an oscillating lever-operated hopper-rubber (also Fig. 300). All these types had grooves from the saddle-quern stage onwards.

The transition from these oscillating forms to the true rotary mill and hand-quern, in which an upper disc-shaped millstone revolves upon a stationary stone, is not at all obvious, though the radial motion of the lever-operated hopper-rubber may have inspired the advance. The general principle is illustrated in Fig. 301; the lower stone is always convex, though it may be

Fig. 300. Reciprocating motion in primitive mills. Left, the ancient Western hopper-rubber, Right, the Olynthian mill, a lever-operated radially oscillating hopper-rubber.

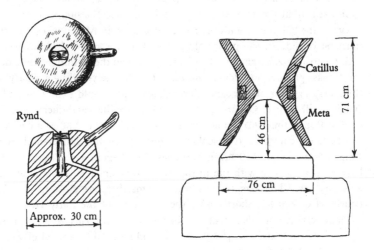

Fig. 301. Rotary mills. Left, the hand-quern or manually operated rotary mill. Right, the Pompeian mill, an animal-powered rotary mill with a large hopper.

only very slightly so, to allow the flour to fall out automatically. The upper is concave with a hole pierced in it through which the grain can fall on to the grinding surfaces. Across this hole is placed a bar – the rynd – which supports the upper stone upon a pin rising from the centre of the lower one. If the pin be continued through the lower stone and attached to a movable lever (the bridge-tree) then a simple method is available for adjusting the clearance between the two stones. This appears in all water-mills and windmills whether in China or the West. Many hand-querns had a handle placed off-centre so as to constitute a crank.

From a structural point of view the rotary mill or quern is the pestle and mortar turned upside down, the pestle being held stationary below, while the mortar with a hole through it, rotates above. This may be another clue in the genesis of the invention. However, the revolving mill is so great an advance on any previous appliance that it must have experienced a decisive act of invention, and the only general precedent it had was the potter's wheel.

In order to compare Chinese evidence with the course of events in Europe, it is necessary first to establish the chronology of the inventions just described. In no Western culture are there any verbs implying the 'turning' of mills before the second century BC, but the contrast between 'push-and-pull mill' and the 'turned mill' is clear from then onwards. Indeed, there is no evidence for rotary mills of any kind in the West until Roman times; saddle-querns were used in all the ancient Western cultures, and the hopper-rubber with a lever – the 'Olynthian mill' named after a town in Macedonia (Fig. 300) – must be considered the principal grain mill of the classical Greek world.

But a great puzzle remains. The hand-quern and the seemingly more developed mass-production 'Pompeian mill' (the 'donkey-mill') – a rotary mill (Fig. 301) – appear at the same time, though one would expect to find the smaller mill earlier. On archaeological evidence the hand-quern goes back only to Pliny's time (AD 70) but mentions in Cato fix 160 BC as a date when it was fairly common. Archaeological finds place the donkey-mill in pre-Roman Spain about 140 BC; nevertheless, the beginnings of rotary milling have been associated with the introduction of commercial bakeries at Rome about 170 BC. The essential invention seems therefore close to the neighbourhood of 200 BC, shortly after the beginning of the Early Han.

There were two words in Chinese for the rotary mill, *mo* (磨 and 䃺) (alternatively *mo* (䃺)) and *long* [*lung*] (礱), depending on the material of which it was made. An ancient dialect word, *chui* [*chhui*] (隤) comprised all these indifferently, and yet another term, *wei* (磑) acquired an even broader significance in that it could include edge-runner mills and roller-mills. The *long* was essentially for hulling grain and especially rice, but the remarkable thing about it was that the Chinese made it from sun-dried or baked clay or wood. When clay was used it was customary to fix teeth of oak and bamboo into the mill-'stones' while still damp (Fig. 302) to act in the same way as grooves incised in stone. The *mo* on the other hand was essentially a stone grist-mill, for grinding husked grain, rice or naked wheat, into flour (see Fig. 303). In all these various types the clearance was adjusted by the height of the central pin bearing. Rotary mills are illustrated frequently in the Chinese agricultural books, and usually shown equipped with a connecting-rod attached to the crank-handle of such length that it could easily be pushed and pulled by several men (cf. Fig. 286, p. 76). This tradition of hand-labour inspired the

Fig. 302. Chinese rotary mills: the *long*, of baked clay or wood, used for taking the husks off grain. The lower bed-stone is here being made. Clay soil is beaten down into the wickerwork holder, and teeth of bamboo (smoked oak for the upper disk), as well as the central pin, are set in place before drying. Diameter of whole structure is about 40 cm.

relevant poem of Lou Shou in his *Geng Zhi Tu* [*Kêng Chih Thu*] (Pictures of Tilling and Weaving):

> Shoulder to shoulder the farmers push and pull,
> Setting the mill in motion with its grinding teeth,
> Making a noise like thunder in the spring.
> As the mill turns the grains fly whirling down,
> Falling in heaps like mountains and rivers,
> Facing each other like high hills.
> They began with a *tou* of grain
> But soon, soon there is plenty to gladden their eyes.

All such uses were current from the beginning of the Song. The real problem is what was happening at the beginning of the Han.

Fig. 303. Typical Chinese hand-quern.

It is first of all clear that the words implying rotary mills do not occur before the beginning of the Qin and Han, though some of them were in use as verbs with the sense of grinding, smoothing and polishing. This at once suggests that variants of the pestle and mortar were the only grain-handling instruments in use in the Zhou and that the rotary mill appeared at the beginning of the second century BC. At this time in China, curiously also, there is a somewhat imperfect agreement between literary and archaeological evidence similar to that which we encountered in the development of rotary milling in Rome.

The point at issue is whether we can feel assured of the existence of rotary mills in the Early Han period. 'Pairs of mills' are mentioned a number of times in the bamboo-slip documents of the Han which have been published, and although few can be exactly dated the whole series runs from 102 BC to AD 93. Particularly interesting is the reference to rotary mills in the *Shi Ben* [*Shih Pen*] (Book of Origins), a text which must certainly be at least as old as the second century BC for Sima Qian [Ssuma Chhien] used it as one of his most important sources in planning and writing the *Shi Ji* [*Shih Chi*] (Historical Records) which he finished in 90 BC. It contains imperial genealogies, explanations of the origin of clan names, and statements regarding inventors legendary and otherwise. What it says is that 'Gongshu Pan [Kungshu Phan] invented the stone (rotary) mill'; this the *Tu Shu Cheng* [*Thu Shu Chi Chhêng*] encyclopaedia of AD 1756 explains in a commentary taken partly from the *Shi Wu Ji Yuan* [*Shih Wu Chi Yuan*] (Records of the Origins of Affairs and Things) of AD 1085. Gongshu Pan is an old friend of ours (p. 20) and cannot be written off as legendary; he was an artisan who flourished within the period 470 to 380 BC. After this we have the *Ji Jiu Pian* [*Chi Chiu Phien*] dictionary of 40 BC, the definitions of the *Fang Yan* (Dictionary of Local Expressions)

Fig. 304. Typical Han tomb model of rotary hand-quern with a double hopper.

of about 15 BC, and then the entry of Xu Shen [Hsü Shen] in the *Shuo Wen Jie Zi* [*Shuo Wên Chieh Tzu*] (Analytical Dictionary of Characters) of AD 100, copied by many subsequent compilers of dictionaries.

As for the archaeological evidence, we have abundant examples of rotary mills with double hoppers from Han tombs in the form of models (Fig. 304 and Fig. 287, p. 77), sometimes alone and sometimes associated with triphammers and other equipment. Although only those specimens from Later Han tombs can as yet be exactly dated, there are far too many of them to warrant the view that none are earlier than the first century AD, and Chinese archaeologists do not hesitate to assign them dates as far back as the Qin.

We thus reach the conclusion that everything points to the first half of the second century BC as the period when rotary mills were in general use – and therefore what was familiar to the Romans Cato and Varro was also familiar to Chao Cuo [Chhao Tsho] and Zhao Guo [Chao Kuo] at the other end of the Old World. This means that we must face a difficult question, diffusion or simultaneous invention? A similar question arises over the approximately simultaneous appearance of the water-mill at the two extremes of East and West. It is a good deal easier, however, to visualise such changes in power source as independent developments, than to accept two quite separate origins for so basic an invention as rotary milling itself – comparable surely with such cultural elements as bronze-founding, the wheel, or cereal agriculture for which no one has been willing to admit independent beginnings. Etruscans and Iberians have been suggested, though there may well have been a common origin about which we still know nothing. The dilemma is even more acute between Chinese and Europeans. We are therefore driven once again to look for some geographically intermediate locality such as Persia (Iran) or Mesopotamia, whence the fundamental discovery could have spread in both directions.

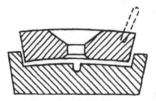

Fig. 305. Pot-quern or paint-mill.

Now an exception to what was said about convexity and concavity is constituted by the so-called 'pot querns' (Fig. 305), in which the upper stone
revolves within the collar of a hollowed cylinder-shaped lower stone. This
design was common enough during the European Middle Ages, though it
seems uncertain whether primarily for grinding grain; some may have been
used for grinding paints. It was never a Chinese type, but some examples from
the Middle East may be very ancient. But these are not the only objects which
make a claim to being millstones of high antiquity in the central region. There
are querns from the neighbourhood of Lake Van in Turkey which, if not
as old as their eighth- or seventh-century BC attribution, might well be old
enough to serve as the archetypes for both the Chinese and the Roman mills.

Dr Needham finds that here a hitherto unsuggested possibility presents
itself. This is the possibility that the Chinese system of making hulling-mills
of baked clay, stoneware or wood might have derived from still earlier
precedents somewhere in or near the Fertile Crescent in Mesopotamia or in
the Indus Valley of Pakistan. In this case all evidence of rotary mills in the
West, as in China, earlier than the second or third centuries BC may well
have perished. If only those which were made of durable stone are caught,
as it were, by the beam of our archaeological torch, their predecessors might
have remained for ever in darkness if the particular agricultural needs of the
Chinese had not induced them to continue through the ages this interesting
method, assuredly cheap, but also perhaps very ancient. And doubtless the
probable locality of these predecessors was neither Etruria nor Spain nor
China, but somewhere in the Middle East. Indeed it is not inconceivable that
only the baked clay mills spread, and that stone ones were independently
developed both by Gongshu Pan and the Etruscans.

Let us turn now to the application to milling of sources of power other than
human muscular strength. The first and simplest improvement of milling
machinery would be the 'mechanical advantage' derived from fitting millstones with gearing, human labour remaining the motive power. This happened quite early in Europe to judge from remains discovered at the Saalburg,
a Roman fortress on the border of South Germany, including two iron shafts
80 cm long and bearing at one end pin-drums (see Fig. 272 *e*, p. 52) with

wooden discs and 'teeth' in the form of iron bars. These lantern gear-wheels were undoubtedly intended for the working of mills by right-angled gearing, and as their location precludes water-power, human or animal-power was their drive. The fortress was abandoned in AD 263, but the gearing repeats exactly a pattern of a Vitruvian water-mill of 25 BC. However, geared hand-mills were a sort of test for inexperienced German engineers of the fifteenth century AD, but in China they are not to be found until modern times, when they have proved useful in the countryside.

The history of the use of gearing with rotary mills and the history of animal-power for them are intimately connected. In the West mills turned by donkeys go back to the very beginning of firm information about rotary mills themselves, for they date from Cato's time (*c.* 160 BC). They precede water-mills,

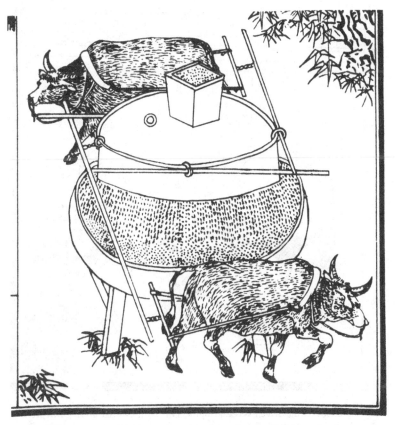

Fig. 306. Ox-driven cereal-grinding mill; from *The Exploitation of the Works of Nature*, AD 1637.

therefore, by over a century, and were probably earlier than querns and hand-mills.

In China, on the other hand, an inversion of the logical order occurs later and seems better established. There water-mills appear to have come first, for while we can find them already in a high state of perfection in AD 31, we get no mention of animal-driven mills before about AD 175, when Xu Jing [Hsü Ching], afterwards an important official, failed to get any post when young and made his living out of mills turned by horses. In the Tang it was customary to use blindfolded mules. Further evidence for the historical inversion of water-power and animal-power from their logical sequence might be drawn from the the curious circumstance that in the time of Wang Zhen [Wang Chên] (AD 1313), and certainly later periods also, mills driven by

Fig. 307. Mule-driven cereal-grinding mill with crossed driving-belt from a larger whim wheel giving a magnification of angular velocity. From *The Exploitation of the Works of Nature*, AD 1637.

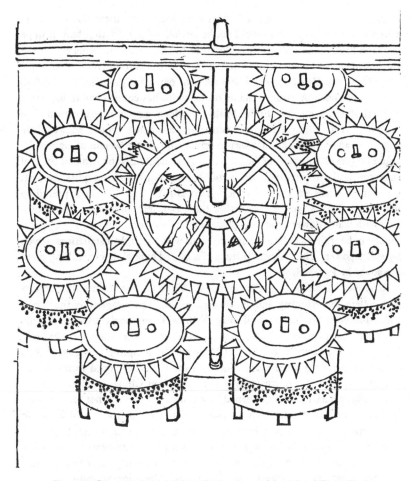

Fig. 308. Geared animal-driven milling plant with eight mills worked directly by the central whim wheel; from the *Treatise on Agriculture*, of AD 1313.

animals were called 'dry water-mill'. In 1360 we read that the mills of the imperial palace were located in the upper storey of a special building, with the plodding donkeys and the gossiping loafers below; this mill-house had been built by an ingenious artisan named Chu. The same plan with the addition of mechanised flour-sifters, is now widely found in Chinese rural areas. In the agricultural treatises of the great tradition we find animal-power applied to mills either directly (Fig. 306) or by means of a driving-belt (Fig. 307).

Once forces greater than man's own were harnessed, there was no reason why a whole series of mills should not be worked off one driving-shaft. Europe seems to have been very slow to realise this, but historians in China give credit to several people in the third century AD for the introduction of multiple geared mills. Ji Han [Chi Han], the eminent naturalist (flourished AD 290 to 307), whose book on strange plants of the south has come down to us, wrote a poetical essay on the 'Eight Mills' in which he says: 'My cousin' Liu Jingxuan [Liu Ching-Hsüan], invented a mill-house which showed rare ingenuity and special skill. (It was so arranged) that the weight of eight mills could be turned by only one bullock . . .' Here doubtless the name of the real inventor has been preserved for us, but he may not have been quite the first, for some sources attribute the design to Du Yu [Tu Yü] in a preceding generation (AD 222 to 284). Whether these multiple mills began in connection with water-power or animal-power is thus not sure, but the 'Eight Mills' tended in later times to become synonymous with multiple installations driven by water-power. The classical picture of a geared animal-driven milling plant is shown in Fig. 308.

Reference has already been made (pp. 49, 57) to the use of wheel-like objects and rollers for grinding and milling purposes. The simplest is the longitudinal-travel edge-runner mill (Fig. 309), still commonly used in China, especially by pharmacists and metallurgists, but little known in Europe. Sometimes it is worked by the feet. It would be tempting to regard this as derived directly from the saddle-quern, if the more developed forms of this ancient device were known from Shang and Zhou China. Or perhaps from the vertical rotary grindstone turned by a crank. Yet that was not so. We have, in fact, to turn to the early appearance in China of the rotary disc-knife for jade-cutting – a technique which goes back at least to the beginning of the Han and perhaps to the beginning of the Zhou (thirteenth century BC). Here, in this rotary tool developed for an industry peculiarly Chinese, we may perhaps recognise the ancestor of all Chinese edge-running mills.

A transition exactly similar to that from grain-rubber to quern took place when the edge-runner wheel was made to revolve in a circular path or trough. This type of mill is not often illustrated with a single wheel in Chinese books, and the oldest picture we have of it shows two wheels diametrically opposite each other (Fig. 310), though a common contemporary variant is to mount the two wheels one immediately behind the other in the form of a bogie. The distinction between wheels and rollers is not sharp, and the roller-mill where a circular path is described, may have been a development from the roller harrow, which is ancient in China and, incidentally, there is evidence that the roller harrows of eighteenth-century Europe were derived from it. Such rollers, if used for threshing, would have given rise very naturally to roller-mills since they would be driven round and round the threshing-floor in a

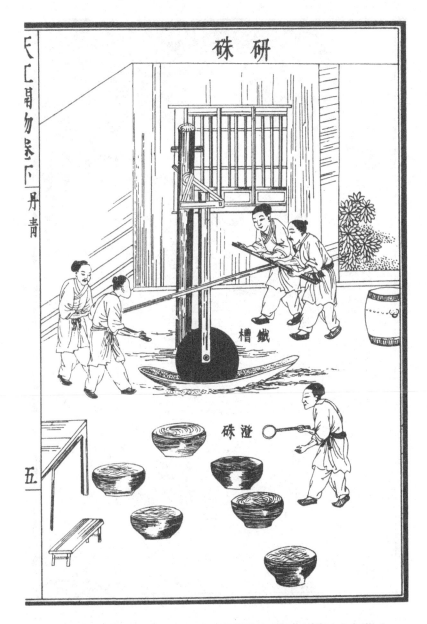

Fig. 309. Longitudinal-travel edge-runner mill, here grinding cinnabar for vermilion, and mainly used in the mineral and pharmaceutical industries; from *The Exploitation of the Works of Nature* of AD 1637.

石輾

石

Fig. 310. Rotary double edge-runner mill for millet, kaoliang, hemp, etc.; from *The Exploitation of the Works of Nature* of AD 1637.

Fig. 311. Hand-roller worked by two girls; from *The Exploitation of the Works of Nature* of AD 1637.

circle. The simple hand-roller (Fig. 311) has always been used mainly for millet in China, but the roller-mill (which often carries a hopper at the opposite side of the beam) is used as a method of hulling rice (Fig. 312).

As to the antiquity of these forms of milling in China, they were undoubtedly familiar in the Han, since they are defined in the *Tong Su Wen* [*Thung Su Wên*] (Commonly Used Synonyms) of Fu Qian [Fu Chhien] about AD 180. Whether or not they go back, like the rotary grain-mill, to the beginning of the Han, it is hard to say. But there is more than meets the eye in the apparently simple edge-runner mill. It is possible that this device may well have been the model for the most archaic of the Chinese theories of the universe – the Gai Tian [Kai Thien] theory where the heavens were thought to be a dome above a convex Earth, separated by a circular trough – in which case it would have existed at least as early as the Warring States period. Furthermore, if the double-wheel type with the wheels on directly opposite sides (Fig. 310) came into use in the Qin or Early Han, it might have provided a model for the idling wheels to be found in the gearing of the south-pointing carriage of the first century AD onwards (p. 189). We have at least proof of the existence of this type of mill in the Tang because of the entertaining story of a particularly strong-armed artisan named Zhang Fen [Chang Fên], who worked for Buddhist monks around AD 855 and could stop a double-wheeled water-powered edge-runner mill unaided.

Comparison with Europe is made a little difficult because of the existence of another puzzle there, namely the unexpected early appearance of forms of rotary mill much more complicated than the rotary quern or the Pompeian grain-mill. It seems that the special needs of the olive-crop so characteristic of Mediterranean lands created a highly complex combination of edge-runner mill and quern as early as the fifth century BC in Greece. Known as the *trapetum*, this separated the stones from the fruit without crushing them as well, after which the oil was expelled from the pulp by crushing. It is illustrated in Fig. 313. There was also a second type of olive-mill, the *mola olearia*, which was simpler (Fig. 313, right). Here crushing occurred only underneath and not at the sides of the grinding 'wheels'.

After classical times the *trapetum* seems to have died out completely in Europe, though there is something very similar to be found in the oil-mills of India. This has a pestle-shaped crusher carried round by an animal pulling a bar pivoted at the centre of the mill; its origin is unknown. On the other hand the *mola olearia* presumably gave rise to the edge-runner and roller mills in Europe, which appeared during and after the Middle Ages. However, when edge-runners appear in Renaissance books they are quite of the Chinese type. Since the common name for this is the 'gunpowder mill', the possibility presents itself that its Western forms may not all derive directly from the *mola olearia* but rather from the Chinese influence exerted in the early fourteenth

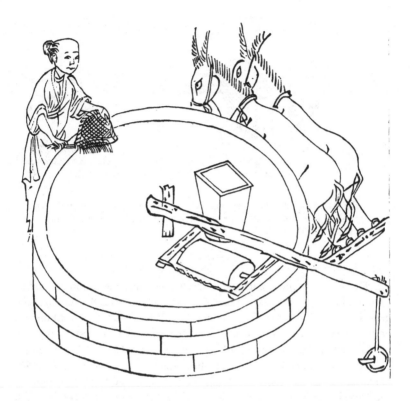

Fig. 312. Animal-driven roller-mill with hopper and road wheel, from the
Treatise on Agriculture of AD 1313.

century at the time of the westward transmission of gunpowder technology.
As for the ultimate origin of the Chinese edge-runner and roller-mills it seems
highly unlikely that they could have derived from a more complex device
developed at the other end of the Old World for an industry quite foreign
to East Asia. Spontaneous evolution from the operations of threshing and
harrowing, or from the rotary disc-knife used in jade-working is much more
probable.

Though the terminology is not good, the roller-mill must be distinguished
from the rolling-mill. While the former consists of one or more rollers travelling continuously in a circular path, the latter utilises the compressing and
shearing properties of two adjacent rollers turning in opposite directions. The
oldest representatives of these mills, which became so important in the Iron

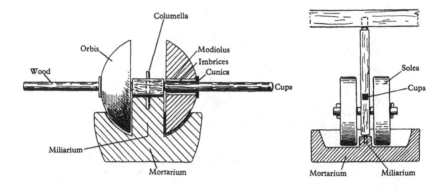

Fig. 313. Left, the *trapetum*, an olive-crushing mill, characteristic of classical Greece. Right, the *mola olearia*, another type of olive-crushing mill, used in the Hellenistic world.

Age for fashioning metal bars and strips, are certainly the cotton-gin and the sugar-mill. The gin is always mounted so as to rotate vertically, the sugar-mill generally moves horizontally. Neither are likely to have been Chinese in origin, since both the cotton plant and the sugar-cane are essentially indigenous to South Asia. Indeed, the gin cannot have been used in China more than a few centuries before its first illustration in the *Nong Shu* [*Nung Shu*] (Treatise on Agriculture) of AD 1313, and the sugar-mill is not illustrated till the *Tiah Gong Kai Wu* [*Thien Kung Khai Wu*] (The Exploitation of the Works of Nature) of AD 1637.

Sifting and pressing

It only remains now to say a few words about the techniques of sifting and pressing. For sifting flour or powders a treadle-operated machine with a rocking motion has long been in use in China. In Fig. 314 the old farmer can be seen throwing his weight from side to side of an oscillating bar to operate a box-sifter. Though Dr Needham has not found any illustration of the treadle machine before the seventeenth century AD, four centuries earlier the *Nong Shu* describes a system of coupling such oscillating sifters to water-wheels with a conversion of rotary motion to motion in a to-and-fro direction (Fig. 315), exactly similar to the rotary to longitudinal action of the Chinese hydraulic metallurgical blowers (to be described in the next volume of this abridgement). The box-sifter must therefore certainly be of the Song and in all probability much older.

The comparative study of pressing plant in East and West presents problems of some interest in the history of technology not hitherto investigated.

Fig. 314. Treadle-operated sifting or bolting machine, from *The Exploitation of the Works of Nature* of AD 1637.

It happens that in this field there is particularly good information about the development of presses for those industries of oil and wine so characteristic of the Mediterranean. We shall therefore proceed from the known to the unknown, first discussing Western methods and then comparing them with the typical presses of Chinese culture.

The basic Hellenistic texts have revealed the types of press sketched in Fig. 316, and inform us that after the first century BC the large presses for olive pulp and grapes were equipped with screw mechanisms rather than the winding gear and weights which had before been used. Although the wedge-press was known, it seems to have been used rather for the preparation of pharmaceutical products, essential oils, cloth and papyrus.

The Chinese pattern was quite different from this. In China the most important type of press has probably always been the one which uses wedges driven home vertically or horizontally with hammers or a suspended battering ram (Fig. 317). This it is at any rate which is described in the *Nong Shu* at the beginning of the fourteenth century. But while the relatively small wedge-presses of Europe were constructed of a framework of beams, the Chinese horizontal oil-presses, used for obtaining the large variety of vegetable oils characteristic of that culture (e.g. soya-bean oil, sesame-seed oil, rape-seed oil, hemp oil, peanut oil), were contrived from great tree-trunks slotted and hollowed out. In these is placed the material to be pressed, made into discs ringed with bamboo rope and bound with straw (see Fig. 316, 4*b*); then the blocks are placed in position and the pressure increased from time to time by the wedges. This method takes advantage of the high tensile strength of the natural wood. It would seem to be ancient, ingenious and without many parallels in the West.

Indirect lever-presses were also much used in traditional Chinese technology. One of these, a rope-clutch press common in the paper and tobacco industries (Fig. 318) is closely similar to type 1*a* in Fig. 316. It differs, however, in having a simple but cunning arrangement of the ropes so that besides hauling down the press beam they act as a brake upon the windlass, thus rendering any pawl and ratchet unnecessary. Again this has all the characteristics of an ancient design. The other main type of Chinese indirect-lever beam-press is distinctly different from any European form (see Fig. 316, 1*c*). Since this is used in a typically Chinese industry, the pressing of soya-bean curd, it is unlikely to have been an importation. Here a weight of stone or iron is used, as in type 1*b*, but instead of being raised by tackle and windlass it is made to depress a lever connected with the beam by an adjustable linkwork arrangement, so that when a certain amount of pressure has been applied the weight may be raised and the perforated ratchet-bar reset so as to continue the pressure as the curd decreases in bulk owing to the expulsion of water.

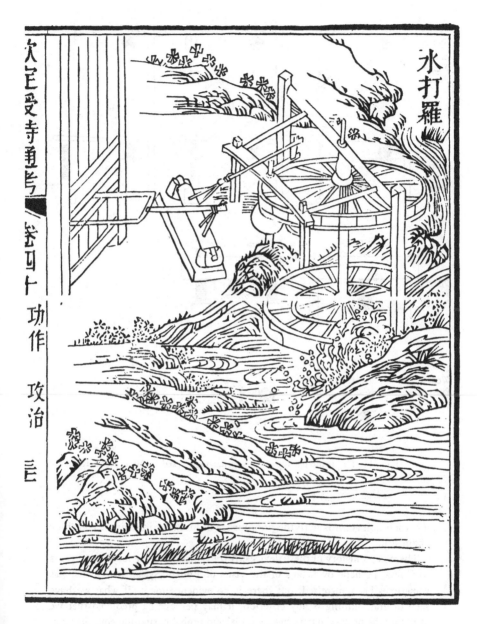

Fig. 315. Hydraulic sifting or bolting machine, constructed on the same principles as the metallurgical blowing-engines. From the *Shou Shi Tong Kao* (Complete Investigation of the Works and Days) of AD 1742.

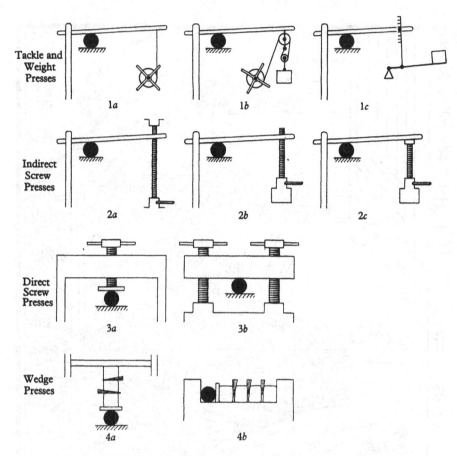

Fig. 316. The principal types of pressing plant used in Greece and Roman Italy.

This solution accords with the prominence of linkwork in Chinese eotechnic engineering.

There is no history of these devices. A transmission of the lever-beam press to China, perhaps by means of the visits of Roman Syrian merchants, has sometimes been surmised, but there is no evidence for it. If what we have seen of grain-milling is any criterion, it may perhaps be predicted that we shall find the presses to have been parallel and probably simultaneous developments deriving from the primitive use of heavy stones to weight the end of a beam, or wedges to tighten discrete objects. The outstanding difference between the two ends of the Old World was the absence of screw-presses from

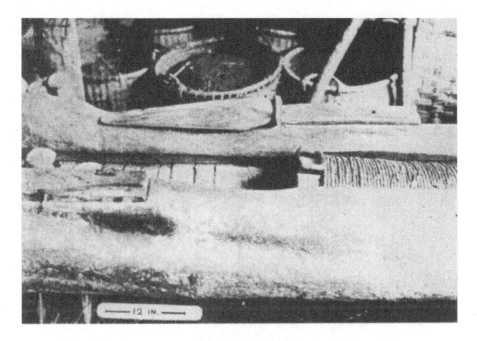

Fig. 317. The most characteristic Chinese oil-press, made from a great tree trunk slotted and hollowed out. The total length of trunk shown is 1.6 m.

China, but this is only another manifestation of the fact that this basic mechanism was foreign to that culture.

PALEOTECHNIC MACHINERY; JESUIT NOVELTY AND REDUNDANCE

So far we have been dealing with machines which were unquestionably and traditionally Chinese. But the Paleotechnic Age of coal and iron exerted its influence on Chinese technology in three engineering books produced there in the early years of the seventeenth century AD by Jesuits or under their influence. These were the *Tai Xi Shui Fa* [*Thai Hsi Shui Fa*] (Hydraulic Machinery of the West) of 1612, the *Qi Qi Tu Shuo* [*Chhi Chhi Thu Shuo*] (Diagrams and Explanations of Wonderful Machines) and the *Zhu Qi Tu Shuo* [*Chu Chhi Thu Shuo*] (Diagrams and Explanations of a Number of Machines), both of 1627.

Many erroneous conclusions have been drawn as to the extent to which they brought to China real novelties in engineering science. Examination has shown that they contain pictures badly redrawn by artists who did not well

Fig. 318. Model of the rope-clutch press as used in the tobacco and paper industries (Mercer Museum). The upper press-beam is forced down upon the lower by a rope made fast to a projecting peg in the windlass between the forked ends of the lower beam. As it also passes in two bights round the windlass outside the forked ends, it exerts a strong braking as well as compressing action, so that the windlass does not tend to fly backward when the hand-spikes are shifted from one hole to another, and a pawl-and-ratchet device is unnecessary.

understand what they were depicting, of illustrations in noted European books, such as the *Théâtre des Instruments Mathématiques et Méchaniques* (*Theatre of Mathematical and Mechanical Instruments*) of 1578 by Jacques Besson, *Diversi e Artificiose Machine* (*Various Artificial Machines*) of 1588 by Agostino Ramelli, the *Novo Teatro di Machini e Edificii* (*New Theatre of Machines and Buildings*) by Vittorio Zonca (1607 and 1621), and the *Machinae Novae* (*New Machines*) of Faustus Verantius which came out in 1615. In other words, a considerable number of the machines and devices which the Jesuits and their collaborators described were specifications which had only just been published in the West. But there were also a number of machines which had been known for a long time previously in Europe, since they can be traced to manuscripts from the fifteenth and even to the thirteenth centuries (see Table 49).

The relevant question therefore arises as to how far the 'novelties' of the Jesuits were new to China. For instance, in the *Rong Zheng Dian* [*Jung Chêng Tien*] (The Great Encyclopaedia) of 1726 pictures of paddle-wheel boats were obviously copied from Western sources. But treadmill-operated paddle-boats were used in China as early as the eighth century AD. Indeed, the demonstration that a certain picture in one of the Jesuit books was directly copied from a Western work does not of itself prove that the idea was new to the Chinese, though it may have been so for the particular individuals with whom the Jesuits were in contact. Take the case of the endless-chain conveyor or excavator. In view of the antiquity of the radial-treadle square-pallet chain-pump in China, it would be almost curious if it should never have occurred to anyone there that the same device, modified, could be used in excavations for earthworks. And indeed, in the late eleventh century AD *Dong Xian Bi Lu* [*Tung Hsien Pi Lu*] (Jottings from the Eastern Side-Hall) Wei Tai [Wei Thai] tells us the following story:

> (In the Xining [Hsi-Ning] reign-period, AD 1068 to 1077) the city of Linzhou [Linchow] had no wells inside the city walls, and it was necessary to draw water from sandy springs outside . . . There was a great desire to extend the city walls to that place (so as to protect the water supply) but the ground was unsafe, and anything built on it was liable to fall down . . .
>
> So the Acting Commissioner Deng Ziqiao [Teng Tzu-Chhiao] said to General Lu Gongbi [Lu Kung-Pi], the commander of the forces in Hedong [Ho-Tung]: 'Formerly there used to be a Ba Zhu Fa [Pa Chu Fa] (lit. pulling-forth axle method). The sands should be shovelled and pulled out and the space filled up with powdered charcoal and cement. City walls can then be built upon this (foundation) without fear of collapse. I should like to use this method and to build new city walls enclosing and protecting the water supply so that Linchow will always be defensible.' Lu Gongbi adopted the suggestion and the plan was carried out. And the walls have remained firm to this day, without any subsidences, so that the New Qin [Chhin] district can be securely defended.

The expression *ba zhu fa* here may well have been figurative, signifying simply an exchange method, in which some sort of cement or concrete was substituted for the sand, but if we prefer to take it more literally we must surmise a continuous-chain excavator on the same plan as the chain-pump but

Table 49. *Some machines described and illustrated in the Chinese books produced under Jesuit influence, yet previously known in China*

NOTE. The sixteenth-century books mentioned here also contained material new to China. A full list of the newly introduced and the already known is to be found in the more complete table (Table 58) in Joseph Needham's *Science and Civilisation in China*, volume 4, part II.

Illustration	Remarks
Qi Qi Tu Shuo [*Chhi Chhi Thu Shuo*] AD1627). All pictures redrawn for *Tu Shu Ji Cheng* [*Thu Shu Chi Chhêng*], *Kao Gongtian* [*Khao kung tien*], ch. 249 *Ji Zhong* [Chi Chung] section (Weight Raising)	
(1 and 2) Steelyard principle	Known in China from the fourth century BC onwards
(3) Windlass (with handspikes) as crane with three-legged derrick	
(4) Windlass (with crank) as crane with four-legged derrick	Not illustrated in traditional Chinese books, but must have been known and used from Han onwards at least. Cf. Fig. 275
(5) Capstan and pulley as crane with four-legged derrick	
(6) As (4) but doubled	
(7) Crank and driving-belt (using mechanical advantage) for crane	
(8) Conveyor; an endless chain of baskets (ordinary crank seen in use, but worm gear shown)	Seems to have been used in Song
(9) Conveyor; an endless chain of boxes (ordinary crank turning the upper sprocket-wheel, as in 8)	
Yin Zhong [Yin Chung] section (Weight Hauling)	
(10) Haulage over rollers by chain running over sprocket-wheel, operated by crank and right-angle gear (mechanical advantage)	With regard to the use of the endless chains above, and the chain purchase here, the chain drives of Su Song in the Song should be remembered
Zhuan Mo [Chuan Mo] section (Grinding-Mills)	
(11) Field mill (two mills mounted on a four-wheeled wagon), turned by animals	Mills mounted on wagons go back to the early Middle Ages in China. See Fig. 331
Zhuan Dui [Chuan Tui] section (Trip-Hammers; Stamp-Mills)	
(12) Stamp-mill, the lugs on the main shaft raising vertical pestles, the shaft turned by two men with cranks	The only difference here from the water-driven trip-hammers used for so many centuries in China was that

Table 49. *contd*

Illustration	Remarks
	the stamping pestles were mounted vertically
Zhu Qi Tu Shuo [*Chu Chhi Thu Shuo*] (AD 1627)	
(1) Mill operated directly by horizontal windmill with vanes or sails of plaited bamboo	This recalls the Chinese windmills for saltworks
(2) Parts for a crossbow trigger	Old in China

using baskets or buckets into which the spoil was shovelled below and ejected above. Perhaps it was something like the endless-belt transporters locally constructed and so much used in China today.

Before summarising the situation concerning transmission in the seventeenth century AD, let us glance briefly at the sections into which part of the *Qi Qi Tu Shuo* [*Chhi Chhi Thu Shuo*] (Diagrams and Explanations of Wonderful Machines) was divided. There was 'Weight Raising', mostly various forms of pulley tackle. This includes the excavators just mentioned, but the curious thing about it is that the steelyard and the crane-windlass should have been carefully explained – instruments which the Chinese had certainly used since the time of the Warring States. The same applies to the windlass in the third section, 'Weight Raising by Turning'. The second section, 'Weight Hauling', had little novelty except the use of the worm. In the fourth, 'Water Lifting', the Archimedean screws and crankshaft pump were no doubt new, but the fifth, 'Grinding-Mills', is devoted mostly to the application of wind-power, a topic not at all new to China. Next there appears a vertical stamp-mill of characteristically European, not Chinese, design; a revolving bookcase which was certainly not European in origin; Alexandrian force-pumps for firefighting; a repeating water-clock with a wheel and ratchet mechanism of Vitruvius, and mechanical cable ploughing.

Cable ploughing would hardly have been much used before the advent of steam power and internal combustion engines. All the same, it is of interest because it seems to be an instance of the adoption in China of the improved gearing arrangements introduced by the Jesuits. Fig. 319 shows the illustration in the *Qi Qi Tu Shuo*; it must have been copied from Besson, though it goes back to the fifteenth century AD in Europe. All these pictures are unrealistic, for they show only windlasses with handspikes, whereas any effective system would require gearing. Such an arrangement is indeed described in AD 1780 by Li Tiaoyuan [Li Thiao-Yuan] in his book *Nan Yue Bi Ji* [*Nan Yüeh Pi Chi*] (Memoirs of the South) as currently in use at that time in Guangdong [Kuangtung].

The 'wooden ox' (*mu niu*) method is a way of ploughing without animals. Two frameworks shaped like the character *ren [jen]* (人) are set up, having inside them certain pulleys fixed, and round these are wound long cables some eighteen metres in length, which are attached by iron pulling-hooks to the plough. When the method is in use, one man guides the plough while two men sit facing each other on (each of) the frameworks. When the turning is in one direction the plough comes forward; when it is in the other the plough goes back. The work which one man can thus perform is equivalent to the strength of two bullocks. This kind of ploughing is considered very good.

The use of the word 'sitting' here suggests that the men were pedalling rather than turning cranks by hand; if so, this was a very Chinese adaption. The power stated also indicates some simple form of gearing. The absence of

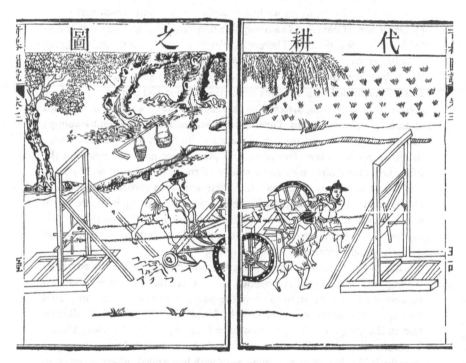

Fig. 319. Mechanical, or cable, ploughing; an illustration from the *Diagrams and Explanations of Wonderful Machines* of AD 1627. The idea was old in Europe, but perhaps not practised either there or in China until after the latter part of the sixteenth century AD.

mechanical advantage in the European fifteenth-century AD drawings suggests that cable ploughing was then only an idea, but it was probably actually employed by the end of the sixteenth century.

Another clear instance of the acceptance of Western techniques by the Chinese in the Jesuit period was the use of force pumps on wheeled vehicles for use as fire engines. These were already being made by Bo Yu [Po Yü] at Suzhou [Suchow] about AD 1635.

A provisional balance sheet of transmissions

Let us now strike a balance-sheet of the position. First we may list ten machines and devices which were assuredly European introductions: (i) Archimedean screw and worm gear, (ii) the double force-pump of Ctesibius of Alexandria, (iii) the Roman drum-treadmill, (iv) the tower-type vertical windmill, (v) the crank-shaft, (vi) the inclined treadmill, (vii) the 'flume-beamed swape' – a long spouted water-raising device, (viii) a water-raising device (swape) worked by a rotating cone-shaped cam' (ix) a swape with pans and (x) the rotary water-pump. Of these by far the most important was the fifth, though the days of its employment in external or internal combustion engines were yet far off. The first, second and tenth were also important in principle though only the first was fundamentally new since, as we have seen, the single-cylinder double-acting force-pump had been familiar in the Song (p. 93 above), and rotary air-compressors (p. 77 above) much earlier. The Archimedean screw for raising water found considerable use, and even some place in literature, in China and Japan during and after the seventeenth century, and the force-pump with its two cylinders spread even more widely as a fire-engine in cities. There was also the introduction of the weight-drive and spring-drive for mechanical clocks, though these only penetrated China later in the seventeenth century. Of the others, some were hardly even practical and never got adopted anywhere.

The seventeenth-century books next show thirteen machines or devices which it was utterly superfluous to consider introducing to China. These are (i) the steelyard, (ii) the windlass, winch and capstan, (iii) the crank, (iv) the pulley, (v) the derrick, (vi) gear-wheels, (vii) the siphon, (viii) chain-pumps, (ix) driving-belts and chain-drives transmitting power, (x) the mobile wagon-mill for grinding 'on the march', (xi) the windmill, (xii) the trip-hammer stamp-mill and (xiii) pivoted devices like the revolving bookcase. All these China had long known and used. The position of three other items – the saw-mill, the suction-lift pump, and the repeating water-clock – is uncertain: the saw-mill because it is not illustrated in traditional books and receives but one mention in them; the suction-lift pump because the buckets in the brine-wells of Sichuan virtually amount to this, so we may suppose the principle at least may have been known to the Chinese; the repeating water-clock, the third, because it was very probably employed in medieval China (see pp. 238–243 below).

Finally, there were, of course, many devices which were not shown in the Jesuit books but which had reached Europe from China at earlier times, or were still to be transferred there. Though detailed in Table 6, p. 76 of volume 1 of this abridgement, a mention here of the chief devices may be helpful: (i) the square-pallet chain-pump, (ii) the double-acting single-cylinder air pump, (iii) deep borehole drilling technique, (iv) the water-powered edge-runner (gunpowder) mill, (v) water-powered trip-hammers for forges as well as grain-pounding, (vi) rotary fans or air-compressors and, notably, the winnowing machine, (vii) power-transmission by driving-belt and by chain-drive, (viii) mechanical clockwork, (ix) the double helical gear, (x) the horizontal warp loom, (xi) the draw-loom, (xii) silk-reeling machinery, (xiii) silk-twisting and doubling machinery, (xiv) the conversion of rotary to reciprocal motion by combination of eccentric, connecting rod and piston-rod and embodied in the water-powered metallurgical blowing engine, (xv) the wheelbarrow, (xvi) the sailing-carriage, (xvii) a form of truss construction in the dishing of vehicle wheels, (xviii) linkwork involved in the trace-harness for horses, (xix) the collar harness, (xx) the crossbow and its companion, the arcuballista, (xxi) the trebuchet or throwing machine using a long pivoted arm, (xxii) rocket flight, (xxiii) the segmental arch bridge, (xxiv) canal lock-gates and (xxv) numerous inventions in naval engineering including the stern-post rudder, watertight compartments and the fore-and-aft rig. Of some of these the Jesuits said nothing, but it was probably due to them that the rotary-fan winnowing machine travelled west at the beginning of the eighteenth century.

There remain a number of further categories of machines and devices about which a word must be said. Some were common the length and breadth of the Old World, and as yet nothing is known of their real first place of origin – among these one might mention the slipways on canals, the most important lock-and-key mechanisms, and the early forms of roller-bearings. Puzzling situations arise when the beginnings of techniques can be dated at closely similar times both in China and the West; the outstanding case of this perhaps is the development of rotary milling at the beginning of the second century BC, and the application of water-power to milling and other purposes during the first century BC. Another category might be constituted by the inventions of which the origin is still very obscure, as in the case of the crank, where we can only say that China takes an honourable place in its history. Then there may be certain Chinese inventions which had died out by the time the Jesuits came. The application of water-power to the slow rotation of astronomical instruments and thence to mechanical clocks was fairly clearly in a period of decline in AD 1600, and another more extreme case might be found in the case of the differential gear, almost certainly used in the south-pointing carriages of the Middle Ages but always confined to court technicians and by the end of the sixteenth century quite lost. In any case, our balance sheet suggests that when traditional Chinese technology came into confrontation with that of the aspiring Renaissance West, it had very little to be ashamed of so far as fundamental principles were concerned. The Jesuits were right inasmuch as they sensed the coming typhoon of Western mechanisation and scientific industry, for which Chinese society was in no way prepared, but standing at the close of the Middle Ages they greatly over-valued the contributions of the Europe of the past.

The steam turbine in the forbidden city

The supreme development of the paleotechnic period was the steam turbine; the Neotechnic Age evolved the automobile; and one scene was set in China. About AD 1671 Father Philippe-Marie Grimaldi (Min Mingwo) organised an elaborate scientific display for the young Kangxi [Khang-Hsi] emperor, and among the optical and pneumatic curiosities which were exhibited on that occasion there figured both a model steam-carriage and a model steam-boat. The carriage was over 60 cm long. The steam issued in a jet and impinged on the vanes of a small wheel which was connected by gearing to a large wheel which rested on the ground. With a second, smaller wheel the carriage travelled in a circle and is reported to have done so for two whole hours. A vehicle similar to this has been reconstructed (Fig. 320). The boat seems to have been a four-wheeled paddle boat.

It appears that experimental models of this kind had been made first by another Jesuit, Father Ferdinand Verbiest about 1665. The customary opinion is that these derived from suggestions for steam-turbines by the Italian engineer Giovanni Branca (flourished 1629), but the difference was that none of these would have been at all practicable, while there is no reason to doubt that the machines of Grimaldi at least worked. It is interesting to note that the turbine principle was not successfully applied to full-scale locomotives till the work of Berger Ljungström in 1922, nor to ship propulsion till about 1897 with the steam-turbines of Sir Charles Parsons. It has been suggested that part of the inspiration of Verbiest and Grimaldi may have been derived from the old Chinese hot-air zoetropes (volume 2 of this abridgement, pp. 361 ff). However, Dr Needham is rather of the opinion that the inclined vanes of the zoetropes descended to the Chinese helicopter top and hence to the airscrew rather than to turbine rotors for steam jets, such as the Jesuits used. Moreover, he thinks that the shape of the boiler suggests another parallel, that of the steam-jet fire-blower. The simplest of such blowers were kettles or boilers with a pinhole so arranged that a jet of steam could be directed on a fire. An incense burner kept glowing by a steam jet is accredited to Philon of

Fig. 320. Model of the steam-turbine road-carriage constructed by the Jesuit Grimaldi, together with many other scientific and technical demonstrations, for the entertainment and instruction of the young Kangxi emperor about AD 1671. A reconstruction in the Museo Naz. della Scienza e della Technica, Milan.

Byzantium about 210 BC; Vitruvius and Heron both mention similar devices. Then in the thirteenth century AD Albertus Magnus describes a device of this kind in detail, and in later technological manuscripts there are several pictures of bronze heads which direct steam jets from their mouths. The jet 'turbine' follows shortly afterwards with Branca and others.

We are now recognisably in the presence of the 'pre-natal' form of the steam-engine's boiler. But in 1629 Branca used an 'Aeolian head' from the mouth of which a steam jet issued, and Athanasius Kircher did the same in 1641. The question was on or into what should the jet be directed? However, in 1601 Giambattista della Porta demonstrated how to empty water from a closed vessel by injecting a jet of steam into it, and then drawing up water by the vacuum left in the vessel after the steam had condensed. In this he was much more truly on the rails which were to lead to the primary break-through of steam power, i.e. the reciprocating steam-engine rather than the steam-turbine.

But here comes the unexpected. The Aeolian heads and busts were a product of European fancy, but for the steam-jet fire-blowers there was another focal area, namely the Himalayan region, especially Tibet and Nepal. There they still take the form of bottle-shaped conical copper kettles sur-mounted by birds' heads, the beaks of which, sometimes quite elongated, point downwards and have the pinhole at the tip. The fact that the ordinary kettle, which emits steam at very little pressure, does not have the same effect, might plead for a single point of origin of the discovery. Certainly, it would have been useful for the dung fires used by travellers on the Old Silk Road and other desert and mountain regions of Central Asia where wood is scarce and altitude considerable. But if so, the remarkable fact presents itself that the Central Asian region of the Mongol-Tibetan-Persian culture is that in which the windmill also seems to have its oldest home. Once again there were Central-Asian slaves in Renaissance Italy whose steam jets may have rein-forced those of Hellenic times. Here at any rate is a suggestive juxtaposition of facts, which might prompt us to look for an ancestor of Branca's proposal in the Arabic-Iranian direction, a first combination of steam-jet and vaned wheel having occurred somewhere in western Central Asia.

THE CARDAN SUSPENSION

There is one device of which it has not been convenient so far to say anything, either among basic principles or early Chinese machines, namely, the gimbals, that seemingly simple combination of rings whereby an object may be main-tained in a horizontal position. Many people are familiar with it because they have seen it in use for one of its most widespread Renaissance applications, the mounting of the mariner's compass to make it independent of the motion of the ship. If three concentric rings are connected together by a series of pivots so that the axes of the pivots are alternately at right angles to each other, and if the central object is weighted and freely movable on the innermost pivot axis, then whatever the position which the outer case comes to occupy the central object will adjust itself so as to maintain its original position.

This device is also known as the Cardan suspension because it was described by Jerome Cardan in his *De subtilitate rerum* (The Subtlety of Things) of AD 1550. But Cardan did not claim the invention as his own; he described a chair on which an emperor could sit without being jolted, and he said that the contrivance had previously been used for oil-lamps. Indeed it had been well known in Europe long before his time, probably taking the use back to the ninth century AD. It was also familiar to the Arabs.

Dr Needham, strolling in a Persian market in 1950, came across two Tibetan brass globe-lamps with gimbal supports and, though they were possibly of recent manufacture, he bought one (Fig. 321). Examples of gimbal-supported incense-burners from China are in the Victoria and Albert Museum in London. Chinese bed-warming braziers with gimbal supports are also well known, while in some provinces lanterns with such a suspension are quite commonly constructed of bamboo, especially those which represent the moon-pearl and are flourished in front of dragons in processions (Fig. 322).

It becomes very unlikely that this Sino-Tibetan tradition of gimbals originated from Cardan and the Italian Renaissance when we realise that we possess from the second century AD a Chinese account of the device which is much earlier than anything in Europe or Islam except a dubious passage in a treatise by Philon of Byzantium. The *Xi Jing Za Ji [Hsi Ching Tsa Chi]* (Miscellaneous Records of the Western Capital) says:

> In Chang'an [Chhang-an] there was a very clever mechanician named Ding Huan [Ting Huan] (flourished AD 180). He made 'Lamps which are always Full' with many strange ornamentations, such as seven dragons or five phoenixes, all interspersed with different kinds of lotuses. He also made a 'Perfume Burner for use among Cushions', otherwise known as the 'Bedclothes Censer'. Originally (such devices) had been connected with Fang Feng [Fang Fêng] but afterwards the method had been lost until (Ding) Huan again began to make them. He fashioned a contrivance of rings which could revolve in all the four directions, so that the body of the burner remained constantly level and could be placed among bedclothes and cushions. For this he gained much renown.

It may be remembered that Ding Huan has been met with before in connection with his zoetrope lamp and air-conditioning fan. The Tang incense-

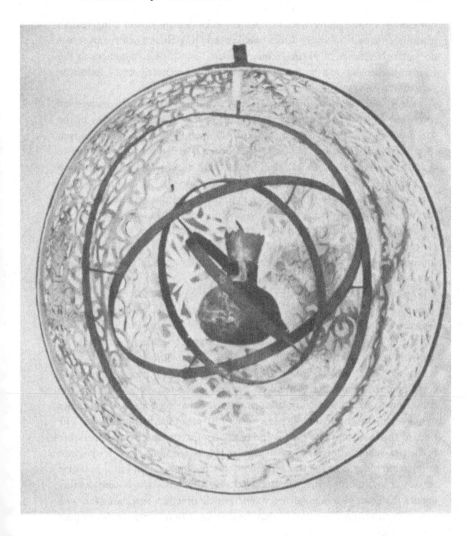

Fig. 321. Interior of a brass Tibetan globe-lamp showing the suspension of four rings and five pivot-axes (photo. Brunney). A candle-end does duty for the original oil-wick.

burners mentioned above would thus seem to be descendants of Ding Huan's unspillable censers in a very direct line. Indeed Ding Huan's achievement was no isolated instance of an ingenious novelty, for Chinese references to the Cardan suspension can be found in the literature of nearly every following period.

Some colour is given to the statement that Ding Huan only revived an invention which had been current long before, by a curious passage in the *Mei Ren Fu* [*Mei Jen Fu*] (Ode on Beautiful Women) written by Sima Xiangru [Ssuma Hsiang-Ju] (died 117 BC). This poem, a defence of his continence against the reproaches of a certain prince whose guest he was, gave occasion for a couple of seduction scenes admirably described. The second of these takes place in an empty palace or remote imperial rest-house, and among the furniture, hangings, bedclothes, etc., carefully described, we find that 'the metal rings (contained the) burning perfume'. The date of composition would be perhaps in the neighbourhood of 140 BC. Thus it is possible that the invention belongs to the second century BC rather than to the second century AD.

It might be said that in the gimbals the effective power-source restoring the original state after displacement is the weight of the object at the centre, and therefore that the power is applied from within outwards. But an invention of even greater importance was made when someone conceived the idea of applying power from the outside, namely the universal joint. Here the outer casing has been changed into a U-piece on the end of a transmitting shaft, corresponding to another U-piece on the end of the receiving shaft, the two being joined by a connecting piece with pivots having axes at right angles to one another. Originally this piece was a ball with pins at right angles. The time of the invention was the seventeenth century AD, and it was due either to Caspar Schott in AD 1644 or to Robert Hooke ten years later. Its greatest application today is in the transmission shafts of automobiles, but it will give service at low speeds when the angle between the two shafts is considerably greater than ever occurs in normal motor-vehicle practice. Few, however, who ride daily in automobiles, realise that the lineage of so important a device goes back to China.

To sum up, therefore, we are faced in this instance with a situation we shall encounter again, in which conclusions as to the origin of an invention are rendered a little difficult by the dubious authenticity of the first European reference. If we adopt the cautious view and regard the reference to gimbals in a work of Philon of Byzantium as a late Arabic insertion into the text, then the credit is Ding Huan's (or Fang Feng's), and it is not unlikely that the role of the Arabs was to transmit the device from further east.

If one should ask about the stimulus which gave rise to the invention of the gimbals, the thought inevitably occurs that it was a derivative from the

Fig. 322. Children engaged in a dragon procession, part of an inlaid lacquer screen in the collection of the late Dr Gwei-Djen Lu (photo. Brunney). In such processions, traditionally customary on the fifteenth day of the first month, a globe-lamp representing the moon-pearl is flourished, as here, in front of the undulating effigy. Gimbals of bamboo within the globe maintain the stability of its light. On the symbolism of the dragon and the pearl, see volume 2 of this abridgement, p. 109 and Fig. 67.

construction of armillary spheres (spheres made of rings) to help astronomers find the position of stars in the sky, since the rings had also to be pivoted within one another. At an earlier stage we caught a glimpse of schools of Chinese artisans making armillary spheres in the first century AD, and it may have been one of these groups to which Ding Huan belonged. At any rate 'Cardan's' suspension is as much Cardan's as 'Pascal's' triangle (volume 1 pp. 54 ff. of this abridgement) is Pascal's.

THE LOCKSMITH'S ART

Together with the millwright, the locksmith was certainly one of those medieval artisans who provided some of the skill at the beginning of the Paleotechnic Age of iron and coal. Unfortunately, there have not been even the barest beginnings of a history of the locksmith's art in Asia, so we can do no more than provide a very simple sketch.

Although this branch of technique elaborated itself in later times into a thousand fanciful complexities, to say nothing of ingenuity devoted to modern time-locks, strong-rooms, and the like, the essentials are simple enough. To begin with there was the bolt, which for convenience was kept attached to the door and made to slide through a block mounted thereon. The first advancement was the addition of a stop and two staples to prevent the bolt from coming out; next an additional staple or lock was added to the wall. Then in order to be able to open the door from outside, a hole was contrived so that the hand could enter, and further refinement diminished this to admit only a mechanical implement, the key. All these types go back to Egyptian antiquity, and have of course lived on in rustic environments until today, both in China and elsewhere.

The 'Homeric' lock – so-called because of its mention in Homer's *Odyssey* – had a bolt drawn to, on leaving, by a chord which passed through the door or door-frame and ended with a loop outside. The kind of key – the *clavicula* – for opening the lock was shaped like a 'clavicle' bone or shoulder-blade. This was essentially a form of crank which, when inserted through a hole, pressed against either a projection or into an indentation on the top or bottom of the bolt. When turned it moved back the bolt, thus opening the door.

The first great invention connected with locks is that of tumblers, i.e. small movable pieces of wood or metal which fell by their own weight so as to engage with 'mortices' or cut-away portions of the bolt, and were then raised by suitable projections on a key. In the beginning they may have been simple pegs or pins hanging on cords and inserted to prevent accidental opening of the door. Possibly the invention was introduced from Egypt to Greece by Theodorus of Samos in the sixth century BC. The projections on the key could work in a number of ways. The key could enter underneath the bolt, push up the tumblers and then slide with the bolt as it was withdrawn; or it could fit into a hole bored along the axis of the bolt itself; or the key could enter above the bolt by a separate opening in the lock-case, the tumblers being [-shaped to receive it. Locks of the first type were used by the Romans (perhaps in the first century AD), and the third is common in China (Fig. 323).

One way of raising tumblers from outside the door was by means of an anchor-shaped key (sometimes called a **T**-key); this was inserted through a vertical slit, given a quarter turn, and then pulled so as to engage with holes in the tumblers; a leverage action would then raise them. Such keys were in use in Rome and in China. In the latter they must be very old, for two examples of Zhou date have been found in old Loyang. This simple pattern many have given rise to locks with 'wards' to prevent any but the correct key from operating the lock, but these were apparently not used in China.

The rotary principle was developed in late Roman times, so that angular displacement now took the place of rectilinear. It seems, indeed, that the ancient *clavicula* may have been adapted from its original use as a bolt-pusher, to become the origin of all rotating keys such as we use today. The Chinese also used, and still use, keys which turn, including quite rustic wooden ones (Fig. 324). Once this principle had been devised, whether for raising tumblers, or for pushing up latches, or for actually sliding the bolt back and forth, the great complexities of cutting keys into so many different shapes answering to different kinds of wards (projecting ridges) were not substantial modifications. Medieval European locks relied mainly on wards, but the use of tumblers was reintroduced towards the end of the eighteenth century AD.

All modern locks, however, made use of what was perhaps the second great invention in the trade, namely the introduction of springs. They dispensed with the necessity of having the tumblers so placed that they must fall under their own weight, for to the tumblers it was usually applied. Springs were used in Roman times in this way, but there were other methods of using them. One of these appears in Fig. 325, which shows one of the kinds of padlock still most widespread in China. When the movable part – the 'bolt' – is fully home, the springs which it carries expand and prevent its removal. The key enters from the other end, and by compressing the springs close to the bolt, permits its withdrawal. However, this padlock was widespread, from Roman Britain, medieval Sweden, Russia, Egypt, Ethiopia, India, Burma and Japan. Often the key had a female screw-thread fitting an invisible male screw on the bolt. As the screw is essentially European, this must have been a transference from West to East.

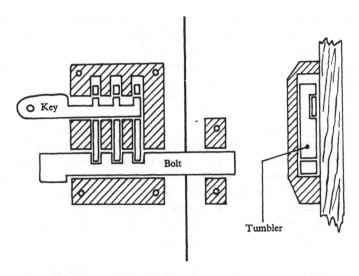

Fig. 323. Mechanism of a type of lock common to Chinese and Roman practice. The key enters above the bolt and parallel with it, lifting the tumblers by mortices made in their sides.

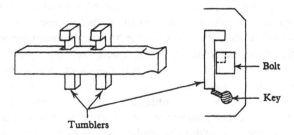

Fig. 324. Mechanism of a common type of Chinese door-lock used for sheds, stores, workshops, etc. The key, which is inserted parallel to the bolt, has two pegs which in their upward path raise L-shaped tumblers notched into the bolt, thereby freeing it for withdrawal. One of the simplest applications of the rotary principle.

So far so good. But the history of the progress of locksmiths' inventions in China remains obscure. One can only offer a few scattered observations. The *Li Ji* [*Li Chi*] (Record of Rites) of AD 80 to 105, may give us Han usage, in which case it seems that some spring mechanism was used. Then the section on punishments in the *Hou Han Shu* (History of the Later Han Dynasty) describes what appears to be a padlock attached to a chain for securing prisoners and operated by an elongated key, thus possibly tying in

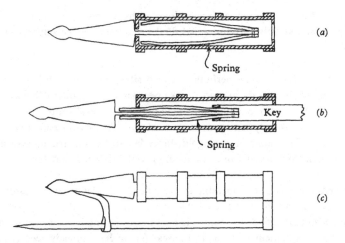

Fig. 325. A type of spring padlock widespread in China. When the bolt is fully home the springs which it bears expand and lock it in position; they are compressed when the bolt is freed by the insertion of a forked key at the opposite end of the padlock.

with a reference in the *Li Ji*. Then in the tenth century AD, Du Guangting [Tu Kuang-Thing] tells us that padlocks were called 'Solomon Seal locks', doubtless because of their resemblance to the tubular stem of this plant (*Polygonatum officinale*), and that they contained metal strips joined together which could be compressed or extended at will. The earliest pictorial representation, however, of the characteristic spring padlocks does not occur until AD 1313 in the *Nong Shu* [*Nung Shu*] (Treatise on Agriculture).

That some keys fitted over projections, and others into holes, is strongly suggested by one of the biographies in the *Nong Shu*, where a 'key and bolt' is mentioned, and the use of a word which primarily means 'male' is interesting in view of the current analogy among locksmiths, engineers and even electricians. In AD 493 we get a glimpse of the Prince of Yulin (Xiao Zhaoye [Hsiao Chao-Yeh]) picking a lock of a city gate with a hooked key. Similarly, the *Yi Yuan* [*I Yuan*] (Garden of Strange Things) says that in the fourth century AD metal keys were as long as 60 cm. While it is hard to deduce much from the Tang term for the lock and key, the *Bei Shi* [*Pei Shih*] (History of the Northern Dynasties) of about AD 670 speaks of 'goose (neck) keys'. One wonders whether some of these were not closely related to the Roman *clavicula* and the Gallo-Roman latch-raiser.

It has been suggested that the tumbler lock and the spring padlock were Asian inventions which reached Europe about the first century BC, but evidence is insufficient. It appears, however, that in the ninth century AD the

Arabs had a high opinion of foreign locksmiths. In his *Examination of Commerce* 'Amr ibn Bahr al-Jáhuz (died 869) listed some of the imports to Iraq as follows:

> From China come perfumes, woven silks, plates and dishes (probably porcelain), paper, ink, peacocks, good spirited horses, saddles, felt, cinnamon and unadulterated 'Greek' rhubarb.
>
> From the Byzantine domains come vessels of gold and silver, *qaisarāanī* coins of pure gold, brocaded stuffs, spirited horses, slave-girls, rare utensils in red copper, inviolable locks, and lyres . . .

This reference is of considerable interest because independent researches in economic history have indicated that it was just during the Tang period in China also that sufficiently unbreakable safes or strong-boxes began to facilitate the development of banking houses. As we have already seen (volume 1, pp. 69 and 70 of this abridgement) relations between China and Byzantium (Fulin) were then particularly close.

In any case, it will be clear from what has been said that there was a rather singular community of pattern between locks and keys throughout the Old World. Whether any technical ideas travelled between Asia and Europe, and if so in what direction, remains an important matter for further investigation. Perhaps the fundamental types developed rather early, in Mesopotamian and Egyptian civilisation, and then spread outwards in all directions, to remain essentially unmodified until modern times.

4

Vehicles for land transport and design of an efficient horse harness

Of the great invention of the wheel, which revolutionised all transportation on land, something has already been said (pp. 43 ff). It is obvious, too, that one of the earliest applications of engineering principles was to vehicles, from Sumerian and ancient Egyptian times onwards. The story of their diffusion and the way they developed would be an epic one if fully told, but this would involve us in the archaeology of times before the earliest Chinese civilisation, and raise questions of their distribution among the races in parts of the Old World furthest removed from China. However, it may now be considered established that the wheeled car, like the potter's wheel, was invented in Sumeria during the Uruk period (c. 3500 to 3200 BC). And it is satisfactory to note that on ancient and medieval Chinese vehicles and their construction there is a substantial literature.

A fundamental distinction may be made between vehicles with a central pole to which animals were yoked at the front end, and vehicles with shafts (prolongations of the side-pieces of the frame) which permitted of more efficient methods of harnessing. There can be no doubt that both types of vehicle derived from sleds or sledges, and that the distinction between single poles and double shafts is more ancient and more important than any which has regard only to the number of wheels. Here it is assumed that the most ancient device was the triangular sled (Fig. 326 a), still used in some parts of the world, and that that gave rise to the rectangular sled (Fig. 326 b); already at this point appears a distinction between the pole and the shafts. Along one line of evolution the triangular sled developed into the triangular cart (Fig. 326 a'), from which it was an easy transition to the pole-chariot (Fig. 326 a''), so familiar from a thousand representations belonging to the antiquity of our own civilisation, and seen indeed on the earliest Mesopotamian saddle-chariots of the fourth millennium BC. This kind of vehicle probably has also some kind of connection with the plough-beam or pole of the plough. The coupling of two of these chariots together would produce a four-wheeled wagon (Fig. 326 e), but this may also have originated from the mounting on

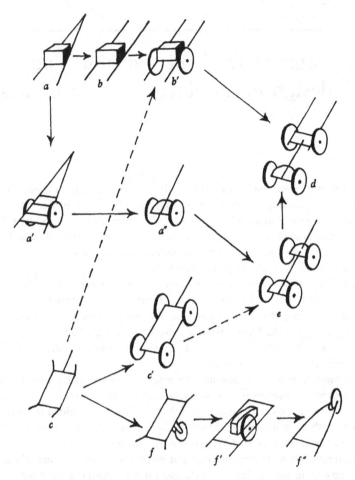

Fig. 326. Diagram to illustrate the evolution of wheeled vehicles. *a*, travois or triangular sled; *a'*, triangular cart; *a"*, pole-chariot; *b*, slide-car or rectangular sled; *b'*, shaft-chariot or true cart; *c*, rectangular draw-sled; *c'*, four-wheeled wagon derived from *c*; *d*, articulated four-wheeled wagon (*b'* + *a"*); *e*, articulated four-wheeled wagon (*a'* + *a"*); *f*, four-handled hod or stretcher mounted on a single central wheel (ceremonial litter, cf. p. 177 below); *f'*, wheelbarrow with central wheel; *f"*, wheelbarrow with forward wheel.

wheels of another kind of rectangular sled (Fig. 326 *c, c'*). Meanwhile the slide-car or rectangular sled with shafts had been mounted on wheels to form the shaft-chariot or true cart (Fig. 326 *b'*). This did not make its appearance in Europe until the end of the Roman Empire, i.e. some half a millennium later than in China. When coupled with the pole chariot, it gave rise in its turn to an articulated four-wheeled wagon (Fig. 326 *d*). The coupling of type *e* is seen on Bronze Age rock carvings in Sweden, and that of type *d* is still found in Northern Italy, while four-wheeled wagons of type *c'* can be noted in La Tène Age bronzes.

The distribution of four-wheeled and two-wheeled vehicles had been the subject of careful study from which it clearly appears that the former were associated with the steppe country of northern and central Asia and Europe, reaching as far west as eastern France and northern Italy, and penetrating into northern India and (formerly) northern China. Everywhere else two-wheeled carts, derived from two-wheeled chariots, held sway. The conclusion can hardly be avoided that this distribution corresponds to the character of the country, two-wheeled vehicles, turning more easily, being associated with obstructions such as hilly roads, hedges and ditches (in Europe), and irrigation channels (in China).

CHARIOTS IN ANCIENT CHINA

That the chariot which the Chinese originally accepted from the Fertile Crescent in the Shang period was of the pole-and-yoke type (Fig. 326 *a"*) seems to be well established from the most ancient forms of the character [*che chhē*] (✦), one of which is shown here. As archaeologists have clearly seen this must imply the use of the inefficient throat-and-girth harness (see below, p. 202). In the Shang and early Zhou periods certain horse trappings found are very similar in design to those of Europe somewhat later (about 3,000 years ago). And the conclusion has been placed beyond all doubt by the discovery in 1950 of a whole park of Warring States chariots in a royal tomb some 80 km south-west of Anyang (Fig. 265, p. 46) These chariots were of the fourth or early third century BC, and though the wood had decayed, its traces in the compacted soil clearly showed the presence of poles and not shafts. Even as late as the Qin (the latter half of the third century BC) there are occasional evidences from bronzes of two-wheeled chariots with poles, and the classical writings speak from time to time of pairs of horses or teams of four. By the fourth century BC, however, shafted vehicles were coming in.

From the Han onwards, all the many representations (see, for instance, Fig. 327) are unanimous that the chariot had shafts (Fig. 326 *b*) and was drawn by one horse only, with an efficient harness. This has been further demonstrated by wooden models of Former Han (first century BC) date excavated from tombs at Changsha [Chhangsha] in 1950. Similar models, in pottery and bronze, all recovered from Han tombs, are to be seen in Chinese museums. Occasionally, moreover (contrary to an often held opinion), horses were marshalled in tandem to draw shafted four-wheeled wagons, as is shown on a Han brick. This would have been impossible without an efficient breast-strap

Fig. 327. A baggage cart from the Yinan [I-nan] tomb reliefs, *c.* AD 193.

harness. In reading Chinese texts, it must be taken into account that the Chinese artisans may well have made little change in their technical terms when, towards the end of the Zhou period, probably at the late Warring States time, they embarked on their great invention of replacing the pole, yoke and inefficient harness of horse chariots, by the shafts and an efficient harness.

Different wheelwrights and wagon builders were detailed in the *Zhou Li* [*Chou Li*] (Record of the Institutions of the Chou Dynasty), and these used technical terms for the parts of wheels. Indeed, so many dimensions are given in the *Kao Gong Ji* [*Khao Kung Chi*] (The Artificers' Record section of the *Zhou Li*) that quotation would be tedious, but it may be interesting to glance at a sketch of the typical Qin and Han vehicle based both on textual and archaeological evidence (Fig. 328): this is noteworthy for the great double curve of the shafts. This was a continuation of a tendency already visible in Sumerian and other Mesopotamian pole types, manifesting the connection which there had certainly been. Frequently the shafts narrowed at the top

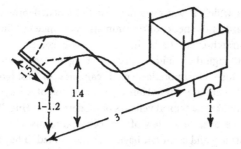

Fig. 328. Sketch of the type of horse drawn chariot or light cart characteristic of the late Warring States, Qin and Han periods, based on the descriptions and specifications in the *Record of the Institutions of the Zhou (Dynasty)*. Dimensions in metres.

to form a small yoke. The general specification (which was modified for special purposes, such as war chariots) provided that the axle, of a definite length or gauge, should be nearly one metre above the ground. The wheel, whether with spokes fitting into a rim with felloes (separate sectors), or solid, was thus almost two metres in diameter. At the meeting of the wheel and axle there was a hollow outer bearing of bronze or iron, around which the hub projected, strengthened with animal sinews glued together and tightly covered over with leather. There could be a metal hub-cap or nave-band, but more usual and more important was the axle-cap which held the wheels in correct position on the axles by means of a linch-pin like our cotter-pins (sometimes known as 'split-pins' since their ends can be splayed out after insertion). The axle-beam supported the lower frame, which carried the body of the vehicle, with its side-pieces, vertical members, and hand-bars. The pole or shafts described a curve 'like water pouring out of a vase'. At the forward end the transverse bar was first attached at its middle by a connection to the pole and carried two 'yokes', but later came (when the change occurred) to unite the two shafts fixedly together.

The text of the *Kao Gong Ji* makes clear that the standard vehicle specifications were closely connected with those for standard weapons such as spears and lances. It says also that if the pole is not sufficiently curved it will weigh upon the horse, while if the curves are too sharp it will break. To understand the significance of this curving as it applied to shafts it is necessary to realise that the traces of the breast-strap harness (Fig. 353, *b* p. 202 below) were attached to the centre point of the shafts between the two inflections (reverse-curve bends), thereby exerting a direct and efficient traction on the vehicles. Traces from any additional horses were connected to it by special rings. The high curving shafts in front of the centre point were really unnecessary, but betray the origin of the arrangement from the earlier pole and yokes, that is to say, from a system in which it was essential that the wooden projection should reach as far forward as the necks of the animals. The ancient Chinese chariot lives on, with straightened shafts, in the common country cart of North China (Fig. 329), though the semi-cylindrical awning or wagon-roof has long replaced the pavilion- or umbrella-shaped canopies of fashionable vehicles of the Han. But there are a few Han representations of military or goods carts extraordinarily like the typical vehicles of modern times (Fig. 330).

Though they seem numerous, only a few of the technical terms which are found in old texts on chariotry and cart-craft have been mentioned. They have considerable interest and importance, for vehicle-building was one of the oldest occupations of artisans, and a knowledge of the terms they coined is indispensable for the study of much later and more complicated engineering texts concerned with milling, textile or timekeeping machinery.

As regards the gauge of Chinese chariots and carts, in volume 1, p. 275 of this abridgement, the standardisation introduced in the third century BC by the first Qin emperor, Shi Huangdi [*Shih Huang Ti*], was mentioned. So lasting was this tradition in particular parts of the country that twenty-two centuries later a traveller found that it was still necessary to change the axle-trees of his carts at one point, since in Shenxi [Shensi], Shanxi [Shansi] and all North-Western China the gauge was some 20 cm broader than that of the eastern provinces, and the size of the ruts made a conversion essential. Stores for this were kept in readiness. This conservatism is outdone only by the carts of the Indus Valley, which in the 1950s possessed the same gauge as their Bronze Age predecessors, to judge by the ruts at Harappa.

Wagons, camp-mills and hand-carts

Before proceeding further, there are a few remarks to be made on certain later Chinese vehicles, some of large size, on imperial carriages noted for their lack of vibration, on camp-mills, i.e. machines for grinding and pounding made mobile by being mounted on vehicles and driven by animal-power, either while moving or when brought to rest.

A very large vehicle was built for the emperor Yang of the Sui dynasty about AD 610. The *Xu Shi Shuo* [*Hsü Shih Shuo*] (Continuation of the *New Discourses on the Talk of the Times*) of AD 1157 says:

> Yuwen Kai [Yüwen Khai] built for Sui Yangdi [Sui Yang Ti] a 'Mobile Wind-Facing Palace'; it carried guards on its upper deck, and there was room for several hundred persons to circulate in it. Below there were wheels and axles, and when pushed along it moved quite easily as if by the help of spirits. Among those who saw it there was no one who was not amazed.

Fig. 329. Country cart at Lanzhou, Ganzhou, 1932.

Fig. 330. Baggage-cart of the Han period; rubbing of a moulded brick from Pengxian, the Chongqing Municipal Museum.

This was probably far from the first time that large vehicles for special purposes had been constructed. Towers on wheels for attacking city walls go back to the treatises on fortification of the followers of Mo Zi [Mo Tzu] (fourth century BC), as also do chariots armoured in various ways.

At the time of the third Tang emperor, Gaozong [Kao Tsung] (AD 650 to 683) and for long afterwards there was something which attracted great attention. Shen Gua [Shen Kua] tells us about it in the *Meng Qi Bi Tan* [*Mêng Chhi Pi Than*] (Dream Pool Essays) thus:

> In the time of Tang Gaozong a large state carriage was made with jade ornaments, and used by him. Three times it was used to carry him to the Tai Shan (for the sacred ceremonies on that venerable mountain), and numerous journeys did he also make in it to distant places, but still today (c. AD 1085) it is sound and firm and steady. If a cup of water is placed in the carriage while it is moving, the water does not spill. In the Jingli [Ching-Li] reign period (AD 1041 to 1048) all the skillful mechanicians available were called upon to construct another such carriage, but it proved too unsteady, so it was left unused. In the Yuanfeng reign period (AD 1078 to 1085) another carriage was built, but in spite of some marvellous workmanship it fell to pieces on its trials before it was ever presented to the emperor. There is still only the Tang vehicle enduring with all its steadiness and freedom from vibrations. But no one has been able to discover the methods by which it was made. Some people maintain that supernatural spirits protect the carriage and bear it along; all I can say is that when one is walking behind it one hears certain vague noises coming from within.

This is surely interesting. Shen Gua was not the man to say that a cup of water would not spill in the carriage if in fact it did, and the story reminds one of the prophecy of George Stephenson so much later about the future of railway transportation. Without going so far as to suggest that the principle of the gimbals was made use of, it may have been that the Tang engineer whose name has not been preserved employed roller-bearings or leaf-springs, though he must have enclosed them in such a way that his successors could not ascertain the technique without taking the equipment to pieces. If it was roller-bearings, he long anticipated Leonardo da Vinci (c. AD 1495), to whom they are attributable in the recent West; and if leaf-springs were used (which in view of their ancient application in the crossbow would seem more probable) the anticipation of European technique would be even more striking, for in the West the earliest certain example is AD 1568. Possibly some kind of suspension was used, with chains or leather straps, as in the *chariots branlants* (swinging carriages) of the European fourteenth century AD.

Much later on we get an echo of giant vehicles in one of the illustrations in the seventeenth-century AD *Tian Gong Kai Wu* [*Thien Kung Khai Wu*] (The Exploitation of the Works of Nature), a four-wheeled dray with nine shafts drawn by eight horses, but the use of such vehicles must have been very restricted by the lack of good roads, if not wholly imaginary. Representations of this kind in paintings of late date showing imperial processions are not uncommon, but the impossibility of the technical details of their harness stamps them as fanciful. Nevertheless, judgement should be reserved on what exactly was constructed in late times in China.

Mills mounted on wagons, which could follow armies like field ovens and field forges, must have been an obvious military requirement from early times. In the West in AD 1607, the engineer Vittorio Zonca drew a picture of a field mill worked by animal-power when the wheels of its carriage were pinned down in camp or near billets. Zonca claimed that it had been invented in AD 1580 by a military engineer, Pompeo Targone. In due course a drawing of Targone's mill found its way to China, for we see it in the *Qi Qi Tu Shuo* [*Chhi Chhi Thu Shuo*] (Diagrams and Explanations of Wonderful Machines) of AD 1627, along with other machines offered by the Jesuit Johann Schreck (Fig. 331), where two mills are worked by gearing from a rotating bar and whippletree harnessed to a single horse.

Presumably neither Schreck nor Wang Zheng knew that just about 1,300 years earlier the history of camp-mills had started in China. In the *Ye Zhong Ji* [*Yeh Chung Chi*] (Record of Affairs at the Capital of the Hunnish Later

Fig. 331. The field mill as it appeared in the *Diagrams and Explanations of Wonderful Machines* of AD 1637. The artist added details not in Zonca's picture, suggesting chain or rope drives from the road wheels; could he have known the earlier Chinese automatic camp mills?

Zhao Dynasty) (*c.* AD 340), Lu Hui, after mentioning that the emperor, Shi Hu [Shih Hu], had a south-pointing carriage and a hodometer (distance measuring vehicle), says:

> He also had pounding-carts (or wagons), mounted on the body of which there were wooden figures pounding all the while with tilt-hammers as the carts moved. Every ten *li* [some five km] traversed meant that one *hu* [about fifty-two litres] of rice was hulled (on each cart). Moreover, he had mill-carts (camp-mills) with rotating millstones mounted on them, and in these also a *hu* of wheat would be ground every ten *li*. All these vehicles were painted red with bright designs. Each was in the charge of one man, and as it moved along all the skill of the construction was displayed. When it stopped the machinery stopped. These mobile mills were made by the Palace Officer, Xie Fei [Hsieh Fei], and the Director of the Imperial Workshops, Wei Mengbian [Wei Mêng-Pien] . . .

Another version has it that each vehicle had one or more hulling tilt-hammers worked off the right wheel or wheels, with rotary millstones worked off the left. Dr Needham has not met with any later accounts of these machines, but it is hard to believe that they died out completely, even if they had been forgotten by Wang Zheng's time. However, the capital, Ye [Yeh] (modern Linchang, near Ahyang), was in northern Henan [Honan], and so in the North China plain, doubtless more suitable for the use of such almost nomadic devices than other centres such as Hangzhou [Hangchow] or Chengdu [Chhêngtu]. Indeed, the invention has the air of a cross-fertilisation between the Hunnish nomadic needs and the Chinese sedentary engineering.

A few words may be added about small vehicles pushed or pulled by humans. That wheeled toys existed in the Han we know from carvings on the wall of a tomb from AD 100, where a child in a long-sleeved gown is propelling a two- or four-wheeled object at the end of a stick. Even more remarkable is a scene in one of the Qianfodong [Chhien-fo-tung] frescoes dating from AD 851, where a woman is pushing a low cradle-shaped vehicle with four wheels. It seems to contain a person lying down, and as it is not quite long enough for an adult, it is probably a baby-carriage or perambulator rather than a hearse or ambulance. As for the shafted hand-cart, it doubtless originated as early as the shafted chariot itself and the special term for it, *nien* (辇) depicts clearly in bronze forms a cart with a pole and yokes pulled by two men. But in its simplest form it must have been much older, for there is a reference in the *Shi Jing* [*Shih Ching*] (Book of Odes) (ninth to fifth centuries BC), and the *Zuo Zhuan* [*Tso Chuan*] (Master Zuoqui's Tradition), describing events of 681 BC, relates how a refugee lord, Nangong Wan [Nankung Wan], pushed his mother in a hand-cart along the road to safety. In the Han such vehicles seem to have been prominently associated with the internal transport of imperial palaces.

All through Chinese history such small two-wheeled vehicles were in use. References to them occur in the *Dong Jing Meng Hua Lu* [*Tung Ching Mêng Hua Lu*] (Dreams of the Glories of the Eastern Capital (Kaifeng)) of about AD 1140. The 'rickshaw' is their late descendant, but was little used in China for some generations until reintroduced from Japan. Several representations are seen on the Dunhuang frescoes. Thus cave no. 431, dating from between the fifth and eighth centuries AD, has a very clear picture of a two-wheeled hand-cart pushed by one man. But in cave no. 148, the Tang paintings of which may be dated at AD 776, there is another man pushing a small cart which might be a single-wheeled wheelbarrow. This brings us to the next phase of our discussion.

The wheelbarrow and the sailing-carriage

Nothing could be more familiar as a vehicle of everyday life than the ordinary wheelbarrow. Perhaps Europeans hardly think of it as a vehicle, since the type used in the West is, as we shall see, very ill-adapted for carrying heavy weights, but the Chinese wheelbarrow is still so constructed that as many as six people may ride on it, and it is universally adopted for freight and passenger transportation. Contrary to what most people would imagine, there is general agreement among historians that the wheelbarrow did not appear in Europe until the late twelfth or even the thirteenth century AD. Builders of castles and cathedrals must have adopted with alacrity a simple device which cut in half the number of labourers required to haul small loads by substituting a wheel for the front man of the hod or stretcher. But from the beginning, the European design placed the wheel at the furthest forward end of the barrow, so that the weight of the burden was distributed equally between the wheel and the man pushing.

That this was not the case with the Chinese wheelbarrow was appreciated by many Europeans who were in China in the seventeenth century AD and afterwards. Thus in 1797 André van Braam Houckgeest wrote:

> Among the carriages employed in this country, is a wheelbarrow,
> singularly constructed, and employed alike for the conveyance of
> persons and goods. According as it is more or less heavily loaded,
> it is directed by one or two persons, the one dragging it after him,
> while the other pushes it forward by the shafts. The wheel, which
> is very large in proportion to the barrow, is placed in the centre of
> the part on which the load is laid, so that the whole weight bears

upon the axle, and barrow men support no part of it, but serve merely to move it forward, and to keep it in equilibrium. The wheel is as it were cased up in a frame made of laths, and covered over with a plank, four or five inches wide. On each side of the barrow is a projection, on which the goods are put, or which serves as a seat for the passengers. A Chinese traveller sits on one side, and thus serves to counter-balance his baggage, which is placed on the other. If his baggage be heavier than himself, it is balanced equally on the two sides, and he seats himself on the board over the wheel, the barrow being purposely contrived to suit such occasions.

The sight of this wheelbarrow thus loaded, was entirely new to me. I could not help remarking its singularity, at the same time that I admired the simplicity of the invention. I even think, that in many cases such a barrow would be found much superior to ours.

In addition to this, I should say that the wheel is at least one metre in diameter, that its spokes are short and numerous, and consequently that the felloes are very deep; and that its convexity on the outer side, instead of being very flat, like common wheels, is of a sharp form. This narrowness of the outer edge (rim) of the wheel appeared to me at first sight very unsuitable. It seemed to me that if broader it would have been better adapted to a clayey soil; but I recollected that at Java, the carts drawn by buffaloes have also wheels with narrow felloes, on purpose that in the rainy season they may cut through strong grounds, in which broad wheels would stick fast – as experience taught the learned M. Hooyman, who attempted to employ broad-wheeled carts in the environs of Batavia, but found himself obliged to follow the custom of the country. I am therefore convinced that the Chinese wheel is the best suited to a clayey soil.

By way of illustration to this passage, Fig. 332 shows the general construction of the Chinese wheelbarrow. The European type may have gained something in controllability, but this cannot have been an important factor, since so many thousands of Chinese wheelbarrows carrying heavy loads still to this day pursue their rapid but squealing courses over the Chinese countryside.

What was not known to van Braam Houckgeest, however, nor to the majority of European historians of technology, is that the date of first appearance of this simple vehicle in China is at least as early as the third century AD. It appears to have been the solution of Zhuge Liang [Chuko Liang], the great general of the Kingdom of Shu so noted for his technical interests, to the problem of supplying his armies. The essential passages occur in the *San Guo Zhi* [*San Kuo Chih*] (History of the Three Kingdoms) of *c.* AD 290:

Fig. 332. The most characteristic type of Chinese wheelbarrow, with central wheel and housing above it, to carry both heavy loads and passengers.

> In the ninth year of the Jianxing [Chien-Hsing] reign-period (AD 231) (Zhuge) Liang again came forth from Qi Shan [Chhi Shan], transporting the army supplies on 'Wooden Oxen'.

A page later, the account continues:

> In the Spring of the twelfth year of the same reign-period (AD 234), knowing that the main army would come out from Yegu [Yeh-ku] (Slanting Valley) he used 'Gliding Horses' to transport supplies.

And furthermore:

> (Zhuge) Liang was a man of great ingenuity. By adding some parts and taking others away, he improved the multiple-bolt arcuballista. Moreover, the 'Wooden Ox' and the 'Gliding Horse' were both invented by him.

Then the commentary of Pei Songzhi [Phei Sung-Chih] (AD 450) on the history goes on to quote the *Wei Shi-Qun Qiu* [*Wei Shih Chhun Chhiu*] (Spring and Summer Annals of the Wei Dynasty) of *c.* AD 360, by Sun Sheng [Sun Shêng], as follows:

The 'Wooden Ox' had a square belly and a curved head, one foot and four legs; its head was compressed into its neck, and its tongue was attached to its belly. It could carry many things, and made thereby the fewer journeys, so it was of the greatest use. It was not suitable for small occasions, but was employed on long journeys; (in one day) it could go several tens of *li* if there was special need, or about twenty *li* if in convoy.... In the time taken by a man (with a similar burden) to go two metres, the 'Wooden Ox' would go six metres. It could carry the food supply (of one man) for a whole year, and yet after twenty *li* the porter would not feel tired.

It is not difficult to discern the wheelbarrow beneath somewhat picturesque phraseology, which almost seems as if it may have been a kind of code, for after all, the design was military, and could well have been considered 'confidential'. Indeed, the whole text points to a wheelbarrow of a type traditional thereafter (Fig. 333). The upright sides and ends would be dismountable, and all its elements can be seen in Fig. 334.

Sometimes there was a disposition to regard Zhuge Liang's devices as mysterious or supernatural, but in the eleventh century AD Gao Cheng [Kao Chhêng] was quite clear that they were Wheelbarrows. He wrote in the *Shi Wu Ji Yuan* [*Shih Wu Chi Yüan*] (Records of the Origins of Affairs and Things) (*c.* AD 1085):

Zhuge Liang, prime minister of Shu, when he took the field, caused to be made the 'Wooden Ox' and 'Gliding Horse' for the

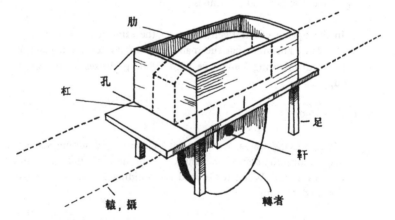

Fig. 333. Reconstruction of Zhuge Liang's army service wheelbarrow according to the specifications in the *Zhuge Liang Ji* (*c.* AD 230) preserved by Sun Sheng (*c.* 360) and Pei Songzhi (*c.* 430).

Fig. 334. The 'one-wheeled push-barrow of the south', from *The Exploitation of the Works of Nature*, AD 1637.

transportation of the army supplies. In Ba [Pa] and Shu (modern Sichuan) the ways were difficult, and these (vehicles) were more convenient for getting over the hills. The 'Wooden Ox' was the small barrow of the present day, and it was so called because it had the shafts projecting in front (so that it was pulled); while the 'Gliding Horse' was the same as that (wheelbarrow) which is pushed by a single person (and so has the shafts projecting behind). Ordinary people nowadays call these 'Jiangzhou [Chiangchow] Barrows'. According to the geographical chapter of the *Ho Han Shu* (History of the Later Han Dynasty) there was a Jiangzhou in Sichuan. At that time Liu Bei [Liu Pei] had occupied the whole of Sichuan, so I suspect that Zhuge Liang's invention was originally made at Jiangzhou, and thus got the name which was continued by later generations.

Gao Cheng thus put forward the plausible suggestion that the 'Wooden Ox' had the shafts pointing forward, while the 'Gliding Horse' had them pointing backwards. The order of the two inventions would then have occurred exactly according to expectation, since the obvious first thought would be to copy the shaft-chariot, and the transposition of the shafts would have taken place a little later, after some practical experience had been gained. In any case, the essence of the invention was economical, for (as we have seen) there is every reason to think that small hand-carts with two wheels, and shafts, for human traction, had been used several centuries beforehand.

One may pause here to point a moral. In the wheelbarrow we have an outstanding example of those many facts which undermine, and indeed over-throw, the classical European stereotype of China as a civilisation with unlimited man-power incapable of inventing and adopting labour-saving devices. Exactly what the economic situation was in the Han when the wheel-barrow first came widely into use remains for further research to elucidate – it may well be that in various historical periods particular parts of China suffered severe labour shortages. In any case long priority is here Chinese, and the surprised and grateful barbarians were European.

In China it has been a tradition to regard Zhuge Liang as the actual inventor of the wheelbarrow. When Dr Needham first drafted this section he also inclined to this opinion, but since then evidence became available which strongly indicates a date rather more than two hundred years earlier as the time of the first invention, i.e. about the end of the Former Han. Of course even in the third century AD others were concerned besides Zhuge Liang. One of the leading technicians of the State of Shu was Pu Yuan [Phu Yüan] whom we shall meet again in connection with iron and steel metallurgy; he held military office in the Western Command under Zhuge and wrote about his men building a 'Wooden Ox'. Another contemporary who may have been concerned was the naturalist and engineer Li Zhuan [Li Chuan] (died *c.* AD 260); like Zhuge Liang (and perhaps under his auspices) he improved the crossbow and the arcuballista. In any event, the wheelbarrow was indissolubly associated in later Chinese song and story with the military exploits of the State of Shu.

The fact is that all these men were at best improvers of a single-wheeled carrier which had been in use throughout the Later Han period. Justification

Fig. 335. Dong Yong's father sitting on a wheelbarrow; a scene from the Wuliang tomb-shrine reliefs of *c.* AD 147. The inscription at the top says 'Yong's father', and on the left 'Dong Yong was a young man from Qiancheng (in northern Shandong)'.

for this view, amounting almost to proof, comes from both carvings and reliefs, and (in a more complex way) from textual sources. As has long been known, the Wu Liang tomb-shrine reliefs, dating from *c.* AD 147, depict the story of Dong Yong [Tung Yung] and his father. Dong Yong, a young man of uncertain date in the Han, became in later ages one of the Twenty-Four Examples of Filial Piety. Losing his mother early he looked after his father, but when the time came he had no funds for the burial, so he borrowed the money on condition that he should be sold into slavery if he could not repay it. From this fate he was redeemed by the remarkable skill of the girl he had married, who after weaving three hundred rolls of silk in a single month revealed that she was in fact the 'Weaving Girl' from on high (the spirit of the star Vega), and disappeared. According to Duan Shi [Tuan Shih], the *Sou Shen Zhi* [*Sou Shen Chih*] (Reports on Spiritual Manifestations) (*c.* AD 348) says that Dong Yong tilled the fields and pushed his father about on a kind of barrow called a *lu che* [*lu chhê*] (鹿車). We shall have more to say about this expression, but the Wu Liang relief does indeed show the father sitting on the shafts of a small vehicle (Fig. 335).

Greater certainty is provided by another carving of about the same date. A wheelbarrow appears sculpted on one of the pillars of the tomb-shrine of

Fig. 336. A wheelbarrow depicted on a moulded brick taken from a tomb of *c.* AD 118 at Chengdu, Sichuan. Only the right-hand bottom portion of the picture is shown. Rubbing from the Chongqing Municipal Museum.

Shen Fujun [Shen Fu-Chün], a Sichuanese notable of about AD 150. Further-more, a Han brick of *c.* AD 118 excavated at Chengdu [Chhêngtu] shows a man pushing a wheelbarrow with a load, in front of a horse-cart with a barrel-vault roof (Fig. 336). Perhaps the load is a dowry box, for this kind of cart is known to have been used by women. Again, as in the tomb-shrine example, the housing of the single wheel is clearly to be seen. Thus the pictorial evidence takes us back to the beginning of the second century AD.

What the texts say is equally important but much more difficult to analyse, and it is important to raise the question of what exactly the *lu che* was. Is the wheelbarrow concealed behind this ancient and soon obsolete term? To answer this the problem essentially consists in determining the nature of what at first sight might be translated 'deer-cart', but which in fact can only be translated 'pulley-barrow'. This entails finding our way through a maze of definitions and long-disused technical terms in ancient dictionaries concerned with various dialects. But first it is necessary to demonstrate the existence and use of this small vehicle and its circumstantial description.

At the beginning of the first century AD it is constantly turning up. The *Hou Han Shu* (History of the Later Han Dynasty) tells us that Bao Xuan [Pao Hsüan], an upright censor who lost his life under Wang Mang in AD 3, married when young Huan Shaojun, [Huan Shao-Chün], the daughter of a wealthy scholar whose disciple he had been. Since Bao's family was poor he was rather embarrassed when the time came to go home, but the excellent girl changed into rough and short clothes, and helped him to push a *lu che* back to his village in the country. This would have been about 30 BC. Again, the same history reports that, fifty years later, when the Red Eyebrows rebel-lion was threatening the Xin [Hsin] and preparing the way for the Later Han, a certain official Zhao Xi [Chao Hsi] and his friends were caught up in it and surrounded. One of them, Han Zhongbo [Han Chung-Po], had just married a beauty and was much afraid that the rebels might harm her. His companions talked of leaving her behind on the road, but he angrily rejected the idea and smearing her face with mud, set her on a *lu che* which he pushed himself. When they met the 'brigands' he said that she was ill, so they all got safely through.

On other occasions, at need, the dead were borne on *lu che*. A virtuous offi-cial, Du Lin [Tu Lin] (died AD 47) pushed a *lu che* at his brother's funeral, and when a friend of Ren Mo [Jen Mo] died at Loyang, he wheeled his corpse many *li* to the tomb. Later the independent-minded Fan Ran [Fan Jan] (died AD 185), when surrounded by enemies, himself pushed his wife on a *lu che* and sent his son out to glean. Many other examples from the Han and some-what later periods could also be given, the term surviving longest naturally enough in poetry. However, it will suffice to quote what Ying Shao says of the *lu che* in his *Feng Su Tong Yi* [*Fêng Su Thung I*] (The Meaning of Popular Traditions and Customs), written in AD 175.

The 'pulley-barrow' is narrow and small, and its design follows that of a pulley-wheel. Some call it the 'carefree barrow'. It may be drawn by an ox or a horse, for which one cuts grass and fodder at stopping-places. Although pushing it is hard work one can lie down when one comes to a rest-house without further worry, hence the name 'carefree barrow'. If you don't have an ox or a horse to attach to it, one man alone can push it wherever he wants to go.

So much for the passing mentions of something which was so familiar that it needed no explanation. When we turn to the ancient lexicographers for further help, we find a vital passage in the *Guang Ya* [*Kuang Ya*] (Enlargement of the Literary Expositor), a dictionary of dialect synonyms compiled by Zhang Yi [Chang I] in the northern State of Wei during the very same years (AD 230 to 232) which saw Zhuge Liang organising his trains of wheelbarrow porters for the supplies of the rival armies of Shu. What the *Guang Ya* says is as follows: 'The *sui che* [*sui chhê*] is also called the *li-lu*. The *dao gui* [*tao kuei*] is also called the *lu che*. Which, being rightly interpreted, is as much as to say: 'The silk-winding (or quilling, or twisting and doubling, i.e. throwing machine) is also in some places called the "pulley-machine". The "rut-maker" is also in some places called the "pulley-barrow".' Good and sufficient grounds for this understanding were provided by Qing [Chhing] scholars such as Wang Niansun [Wang Nien-Sun] in his commentary of AD 1796.

The *Fang Yan* [*Fang Yen*] (Dictionary of Local Expressions) of Yang Xiong [Yang Hsiung] next confirms that at its much earlier date (15 BC) the *sui che* was called the *li-lu che* in the regions of Zhao [Chao] and Wei – another way of saying 'pulley-machine'. That it had to have at least one pulley, rapidly rotated by the belt from the driving wheel, follows from the nature of the case. The basic linguistic difficulty which we are facing is that the word *che* was always ambiguous in that it could mean indifferently a stationary machine or a mobile vehicle. It is as if one should speak of a 'single-wheeler' without giving any clue as to what it did. Moreover, both the components *li* and *lu* could also mean the rut or track left by a vehicle wheel. So returning to the *Guang Ya*'s second statement, we see that the *lu che*, which we know from a mass of other evidence was sometimes a small vehicle, and had nothing whatever to do with deer, could be either a 'pulley-barrow' or a 'pulley-machine', in the first case a 'rut-maker', in the second a 'trace-maker', but in all cases a 'single-wheeler'. Hence what Bao Xuan and his excellent wife were pushing in the last years of the Former Han dynasty was nothing other than the new-fangled device, the wheelbarrow.

Here we could take leave of the matter and go on our way, but the most ancient origin of the words *li-lu* or *lu lu* remains intriguing, and this terminological rubbish heap is worth turning over. In one of the *Shi Jing* [*Shih Ching*] odes a chariot description includes the phrase 'the pole is curved like a roof-

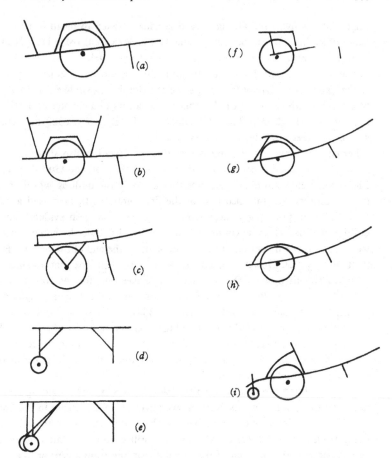

Fig. 337. Types of Chinese wheelbarrows. (*a*) Wheel central, with housing, taking all the load; a substitute for a pack-animal, the 'pack-horse' type. Jiangxi and other provinces. (*b*) The same with wagon sides, identified with the presumptive design of Zhuge Liang. Northern Sichuan. (*c*) Wheel central but centre of gravity high, without housing, axle on sloping struts, flat car sides, used for moving earth. Shenxi. (*d*) Small wheel forward, centre of gravity high, axle on vertical stayed struts, stayed support. Sichuan. (*e*) Two small wheels forward, centre of gravity high, axle on vertical stayed struts, stayed support; (*f*) wheel far forward with rectangular housing, straight frame, the 'half-stretcher' type. Shanxi and many other provinces. This is perhaps the oldest form of the invention, for illustrations take it back to the second century AD, and textual evidence to the first century BC, though the latter is not decisive as to the form. (*g*) Intermediate type with curving frame and curved housing. Western Sichuan. (*h*) Intermediate type with curving frame and streamlined housing. Western Sichuan. (*i*) Intermediate type similar to (*g*) but with a small auxiliary wheel at the front end of the frame, useful for clearing obstacles. Hunan and other provinces.

ridge and has five *wu*. The last word is usually taken to mean ornamental bands of leather, but the ancient commentary of Mao Heng [Mao Hêng] (third or second century BC) explains that the five bands each had a *wu* and that this was in fact a *li-lu*. Here rings to guide the traces or reins must be meant, so that the earliest form of the phrase was the simplest possible kind of pulley-block (*c.* seventh century BC). Another usage occurs in a chapter on fortifications in the *Mo Zi* [*Mo Tzu*] (The Book of Master Mo) (*c.* 320 BC) where a cord and reel are referred to as a *li-lu*.

Perhaps still more interesting is that this investigation also shows us that the *Fang Yan* (Dictionary of Local Expressions) of 15 BC and Guo Po's [Kuo Pho's] commentary provide evidence that not only the quilling wheel of the twelfth century AD, but also that of the first century AD, possessed a belt-drive. Thus in pursuing wheelbarrows we come upon fresh evidence about quilling wheels, which is confirmed in other works. Indeed, the conclusion we come to is the invention of the driving-belt by the end of the Former Han (first century AD). However, from the texts it could just be that a vehicle is intended and then again the wheelbarrow shakes out, for the cord or rope of the belt-drive could also be the sling so often attached to the shafts and passing over the porter's shoulders (see Fig. 334), an aid to stabilisation super-fluous for any two-wheeled vehicle. Thus when we return to the *Guang Ya* we may feel safe in rephrasing it as follows: 'The silk-winding (trace-maker) for the bobbins is also called the "pulley-machine"; the (mobile) rut-maker is also called the "pulley-barrow".'

What the difference was between the 'Wooden Ox' and the 'Gliding Horse' of Zhuge Liang remains unsolved. As we have seen, Gao Cheng in the Song thought that it depended on whether the shafts were before or behind. Certainly the traditional types of wheelbarrow still current in China today are very numerous, the position of the bearings varying from a central to a very forward point. In the most characteristic Chinese design, as shown in the *Tian Gong Kai Wu* (Exploitation of the Works of Nature) of 1637 (Figs 334 & 337 (*a*)), the wheel seems to have been conceived as a substitute for a pack animal and takes all the load. One of these types, common at Guangyuan [Kuang-yuan] (Fig. 337(*b*)) is regarded by Chinese historians as particularly likely to preserve the pattern used by Zhuge Liang. Another (Fig. 337(*c*)) is very much in evidence on the sites of public works. But the wheel-bearings are also often very far forward; Fig. 338 (see also Fig. 337 (*f*)) shows a barrow of this kind – particularly interesting since it resembles so closely the barrow on which Dong Yong's father is sitting in the Wuliang relief (Fig. 335). Often however the wheel-housing is curved rather than rectangular, as in the Sichuanese form shown in Fig. 339 (see also Fig. 337 (*g*), (*h*)), with the wheel less far forward. And the two main types may even be combined (Fig. 337 (*i*)).

Fig. 338. Wheelbarrow of type (*f*) on the road between Xian and Lindong, Shenxi in 1958.

Fig. 339. Wheelbarrow of type (*g*) near Guanxian, Sichuan in 1958.

Of the transmission of the technique to Europe about the beginning of the thirteenth century AD nothing whatever is known, nor is information easily obtained about the wheelbarrow in the geographically intermediate cultures. But the fact that the wheel in the European types is invariably very far forward, as if to replace one man out of two carrying a hod or stretcher, and therefore taking only half the load, may mean that this was another case of 'stimulus diffusion'. Perhaps Westerners simply heard that something of the kind had been done, and proceeded to imitate according to their lights without knowing any exact specifications. At the same time it is not possible to exclude Western influence on contemporary Chinese designs, and one might be

Fig. 340. Korean ceremonial chair-litter; the 'bearers' have simply to guide the vehicle. For its evolutionary position, see Fig. 326.

inclined to see it in the 'half-stretcher' rather than the 'pack-horse' types, though if the Wuliang tomb-shrine is any indication the former was in China from the beginning, and Zhuge Liang's innovation might have been precisely the latter.

Among the curious designs are those which raise the carrying surface high above one (or even two) small but broad-rimmed wheels (Fig. 337 (*d*), (*e*)) with the aid of oblique stays. Those which Dr Needham has seen in Western China always have the wheels well forward, but in neighbouring Korea a ceremonial chair-litter, in which four handles are guided, not carried, by four bearers, and the main weight is taken on a central wheel with struts, stays remarkably reminiscent of aircraft landing-gear (Fig. 340). This is perhaps an archaic intermediate form between the four-handled hod or stretcher (or the draw-sled) and the wheelbarrow proper. What may be a literary reference to this occurs in the *Hua Man Ji* [*Hua Man Chi*] (Painted Walls) of about AD 110, where we read of a litter mounted on a turning wheel. It is noteworthy that the character for litter contains the wheel radical, and indications exist that the wheeled litter may go back to the Former Han. In *Qian Han Shu* [*Chhien Han Shu*] (History of the Former Han Dynasty) there is a biography of Yan Zhu; this describes an expeditionary force sent to Nan Yue [Nan Yueh] in 135 BC, and says 'Chariot-litters (with supplies) passed through the mountains.' This puzzled commentators, but what we probably have here is an interesting, though still very obscure, chapter in the early history of the Chinese wheelbarrow.

A few later references are not without interest. Previously (p. 103 above) mention was made of Ge You [Ko Yu], a semi-legendary personage, Sichuanese of course, who rode away into the mountains on a wooden sheep of his own invention. Dr Needham suggests that this may be a piece of folklore

Fig. 341. Wheelbarrows in the streets of the capital, Kaifeng, in AD 1125 from a painting by Zhang Zeduan 'Returning up the River to the City at the Spring Festival'. On the left an empty wheelbarrow being loaded with sack-like objects outside the best hotel; on the right, another one of large wheel size drawn by a mule and guided by two men who pull and push.

which had the real wheelbarrows of Zhuge Liang for its basis. But the wheelbarrow spread far beyond Sichuan, and in AD 1176 Zeng Minxing [Tsêng Min-Hsing] alluded to the military use of the wheelbarrow in forming protective encampments or 'mobile forts' in Jiangxiang [Chiang-Hsiang]. Not long afterwards Zhu Yi [Chu I] tells us that in the Tang, when Liu Meng [Liu Mêng] went to pacify the region which is now Ningxia [Ninghsia], he relied on the wheelbarrow for the transportation of his supplies.

The mention of Jiangxiang by Zeng Mingxing reminds us of Gao Cheng's [Kao Chheng's] appellation of 'Jiangzhou barrows'. Jiangxiang is a little town in the extreme south-west corner of Sichuan, lying in mountainous regions, and in the sources Ge You is described as a Qiang [Chhiang], i.e. a member of the tribal peoples indigenous to just this region. We shall probably not go far astray then if we accept it as the original home of the invention in the first century BC, and regard Ge You either as the first inventor or perhaps the canonised 'technic deity' of the makers of wheelbarrows.

Animal traction was applied to one-wheel vehicles from an early date in China, as the passage from the *Feng Su Tong Yi* [*Fêng Su Thung I*] is alone sufficient to witness (p. 174 above). Perhaps our best illustrations of this may be seen in the famous painting of Zhang Zeduan [Chang Tsê-Tuan], finished

Fig. 342. A train of intermediate-type wheelbarrows with auxiliary wheels at the front, type (*i*), on one of the paths through the Nanling hills between the Yangzi Valley and Canton.

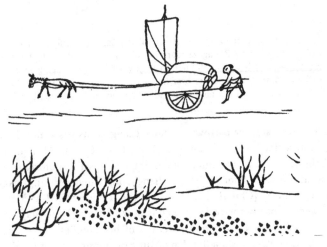

Fig. 343. Sketch of a sailing wheelbarrow, the sail assisting animal traction.

by AD 1126, which depicts popular life of the capital Kaifeng [Khaifêng] at the time of the Spring Festival. Fig. 341 is part of the painting which shows many wheelbarrows moving or stationary in the streets of the city. All but one have the large central wheel and some are very heavily laden; during the loading and unloading they rest on their side legs. One is being pushed by a single man, but in the other cases a porter steadies the vehicle by the shafts behind, while traction is effected either by one man in shafts and one mule or donkey with collar-harness and traces, or by two animals side by side similarly attached.

But perhaps the most important invention was the use of sails, so that wheelbarrows like ships could be borne along by the force of the wind. This admirable device is still used in China, notably in Honan and the coastal provinces such as Shandong. Fig. 343 is a little sketch from Linqing's [Lin-Chhing's] *Hong Xue Yin Yuan Tu Ji* [*Hung Hsüeh Yin Yüan Thu Chi*] (Illustrated Record of Memories of the Events Which Had to Happen in My Life) of AD 1849 which shows a wheelbarrow drawn by an animal which is assisted by a sail. There was also (Fig. 344) a careful diagram published in AD 1797 by van Braam Houckgeest (*An Authentic Account of the Embassy of the Dutch East-India Company to the Court of the Emperor of China in the years 1794 and 1795 . . .*), from which one can see that the sails are typical junk fore-and-aft-slat sails with their multiple ropes. The words of van Braam, too, are worth quoting:

Near the southern border of Shandong one finds a kind of
wheelbarrow much larger than that which I have been describing,
and drawn by a horse or a mule. But judge my surprise when
today I saw a whole fleet of wheelbarrows of the same size. I say,
with deliberation, a fleet, for each of them had a sail, mounted on
a small mast exactly fixed in a socket arranged at the forward end

Fig. 344. Diagram of sailing wheelbarrow showing the batten sail and multiple ropes so characteristic of Chinese practice. From book by van Braam Houckgeest, AD 1797.

of the barrow. The sail, made of matting, or more often of cloth, is 1.5 or 1.8 metres high, and 0.9 or 1.2 metres broad, with stays, sheets (ropes), and halyards, just as on a Chinese ship. The sheets join the shafts of the wheelbarrow and can thus be manipulated by the man in charge.

Unfortunately there are very few references in Chinese literature to sailing-wheelbarrows, so that it is not yet possible to determine the time of their introduction.

The impact which this ingenuity made upon the first European visitors to China in the sixteenth century AD, however, can hardly be imagined. In 1585 Gonzales de Mendoza wrote that the Chinese 'are great inventors of things, and they have amongst them many coches and wagons that goe with sailes, and made such industrie and policie that they do govern them with great ease . . .' Jan van Linschoten said something similar a dozen years later.

These accounts and others caught the imagination of the European map-makers, so that one finds small vignettes of the land-sailing carriages on almost every atlas published in the sixteenth and seventeenth centuries AD where a map of China is provided. And in his *Paradise Lost* Milton immortalised the respect which Europeans could still feel for the strange techniques of China with the words:

> Of Sericana, where Chineses drive
> With sails and wind their cany wagons light . . .

This was in 1665, but the whole country was fascinated by the story, and puzzled interest really did continue long after the late sixteenth century. Thus in Thomas Birch's *History of the Royal Society* of 1756 we find the following somewhat cryptic entry:

> Apr. 1st. 1663; Mr Hooke's paper concerning the Chinese cart with one wheel, mentioned by Martinius in his *Atlas Sinensis*, was read, and discoursed upon, that the said cart was like a wheelbarrow . . .

Then on 23 November of the same year, Mr Robert Hooke displayed 'a paste-board model of his engine with one wheel, to travel in with ease and speed . . .' Thomas Hobbes had also been experimenting. In his *Elements of Philosophy* (1655), which included natural philosophy, he discussed the forces which act upon a sailing ship and illustrated his argument by trials made on a wooden model of a sailing-carriage. Then in 1684 the French Academy of Sciences prepared a questionnaire for the Jesuit Philippe Couplet to take back to China with him, in which details of the sailing-carriages were requested.

But *The Sacrifice*, Sir Francis Fane's tragedy about Tamerlane published in 1686, engagingly ridiculed Chinese antiquity and inventions, including the sailing-carriage. On the other hand Leibniz, when projecting a kind of science museum, suggested that the exhibits should certainly comprise 'le chariot à voiles de Hollande – ou plutost de la Chine'. What Leibniz meant by the sailing-carriages of Holland we shall see in a moment but first, there are one or two rather important early Chinese references at which we must glance.

The emperor Yuan of the Liang, in his *Jin Lou Zi* [*Chin Lou Tzu*] (Book of the Golden Hall Master), wrote 'Gaocang Wushu [Kaotshang Wu-Shu] succeeded in making a wind-driven carriage which could carry thirty men, and in a single day could travel several hundred *li*.' Nothing else is known of this engineer or his sailing-car, but its performance would not be at all impossible, as we shall see shortly. This would have been about AD 550, and then comes around 610 the 'Mobile Wind-Facing Palace' of Yuwen Kai [Yüwen Khai] mentioned in connection with large wagons (p. 160 above). Allowing for some exaggeration in the account, it seems not unlikely that this was also a land-sailing vehicle.

After that there is a long silence, until just on a thousand years later, when Simon Stevin, the great Dutch mathematician and engineer, constructed a sailing-carriage of which we have many historical details. In what was probably the autumn of 1600, Prince Maurice of Nassau invited several ambassadors and distinguished guests to take part in an experiment with a sailing-carriage constructed by Stevin. There were two of these, one larger than the other, and both succeeded in accomplishing the distance between Scheveningen and Petten along the beach in less than two hours though it took fourteen to walk. Fig. 345 shows the vehicles, and it seems probable that

Fig. 345. The successful sailing carriages or 'land-yachts' constructed by Simon Stevin (AD 1600).

in building them Stevin had read the descriptions of Chinese sailing-carriages by van Linschoten and Mendoza. His successful trials were therefore a direct result of the contact of Europeans with China.

Fortunately we do not have to rely on a historical investigation for a report of such sailing-carriages in Europe. There is contemporary seventeenth-century testimony in a voice from the past, that of the celebrated Pierre Gassendi, writing the life of his friend Fabri de Peiresc. Speaking of the year AD 1606, Gassendi says:

> Also he stept aside to Scheveling, to make triall of the carriage and swiftnesse of the waggon, which some yeers before was made with such Art, that it would run swiftly with sails upon the land, as a ship does in the sea. For he had heard how Grave Maurice, after the victory at Nieuport, for triall sake, got up into it, with Don Francisco Mendoza taken in the fight, and within two hours was carried to Putten which was 87 kilometres from Scheveling. He therefore would needs try the same, and was wont to tell us, how he was amazed, when being driven by a very strong gale of wind, yet he perceived it not (for he went as quick as the wind), and when he saw how they flew over the ditches he met with, and skimmed along upon the surface only, of standing waters which were frequently in the way; how men which ran before seemed to run backwards, and how places which seemed an huge way off, were passed by almost in a moment; and some other such like passages.

Long were the reverberations of these results. Stevin's land-ship was copied and recopied during the next couple of hundred years. Indeed, they have continued in the twentieth century in many places from the north coast of Belgium and France and in California. Dr Needham remembers seeing them and riding in one on the beach at La Panne in 1907 when they were being modernised with a tricycle-wheel system, pneumatic tyres and a light tubular construction, and capable of speeds of the fastest express train, especially when sailing upon ice rather than sand.

It might be thought that the sailing-carriage (like the south-pointing carriage – see pp. 189 ff) was merely one of the curiosities of history. But these things have to be placed correctly in the perspective of technological development as a whole. For the sailing-carriage was able to travel with what was, in AD 1600, almost incredible speed. And here is the root of the matter, for we may say that this transmission from Chinese technology (strange as it may seem) was the first to accustom the European mind to the possibility of high-speed transit on land. A distance of nearly 97 km was covered in less than two hours, which must have meant that some stretches were travelled at a speed higher than 50 km per hour, perhaps in the neighbourhood of 65. When one remembers the excitement caused by modest speeds in the first days of railways, one is not inclined to underestimate the impact on European culture of what was really the first essay at rapid transportation. The Chinese stimulus, if it was no better, cannot be ignored, and the results were overwhelming.

THE HODOMETER

The idea of a vehicle which would register the distance traversed appealed to mechanicians in more than one ancient civilisation. The hodometer or 'way-measurer', was quite a simple proposition mechanically. All that was necessary was to make one of the road-wheels drive a system of toothed wheels constituting a reduction-gear train, so that one or more pins revolved slowly, releasing catches at predetermined intervals and striking drums or gongs.

Under the name of 'recording drum carriage', this apparatus is mentioned in most of the official dynastic histories from the Jin onwards. None of these references describe the mechanism, except that in the *Song Shi* [*Sung Shih*] (History of the Song Dynasty), which we shall examine in a moment. Some of them say, however, that at the end of each *li* traversed a wooden figure struck a drum while at the end of each ten *li* another figure struck a gong. If the text of the *Gu Jin Zhu* [*Ku Chin Chu*] (Commentary on Things Old and New) of Cui Bao [Tshui Pao] is to be trusted, this double arrangement existed already in third century AD, but since its genuineness is a little uncerain, it may well be preferable to assume that the more complex machine did not come till the Sui or Tang. The names of certain engineers who constructed such hodometers with conspicuous success have been preserved, notably Jin Gongli [Chin Kung-Li] in the Tang (ninth century AD), Su Bi [Su Pi] in the Wu Dai or Song (tenth or eleventh century AD), Lu Daolong [Lu Tao-Lung] in AD 1027 and Wu Deren [Wu Tê-Jen] in 1107.

On the above evidence, the invention probably dates at least from the time of Ma Zhun [Ma Chun] (flourished AD 220 to 265). This conclusion is strengthened by the fact that the *Sun Zi Suan Jing* [*Sun Tzu Suan Ching*] (Master Sun's Mathematical Manual), which dates from the third to the fifth centuries AD, contains a hodometer problem:

The distance between Chang'an [Chhang-an] and Loyang is 900 *li* Suppose a vehicle the wheel of which covers in one rotation 1 *Zhang* [*chang*] and 8 *chi* [*chhih*] (5.5 metres). How many rotations of the wheel will there be between the two cities?

Apart from the book of Cui Bao, the oldest description is in the *Jin Shu [Chin Shu]* (History of the Jin dynasty) (AD 635) which says:

> The kilometre-measuring drum-carriage is drawn by four horses. Its shape is like that of the south-pointing carriage. In the middle of it there is a wooden figure of a man holding a drumstick in front of a drum. At the completion of every *li*, the figure strikes a blow upon the drum.

Cui Bao's own words (*c.* AD 300) are as follows:

> The *Da Zhang Che [Ta Chang Chhê]* was for knowing the distance along a road. It began in the Western Capital. It is also called the kilometre-measuring carriage. It has two storeys both with a wooden figure. After every *li* traversed the lower figure strikes a drum; after every ten *li* the upper one rings a small bell. The *Shang Fang Gu Shi [Shang Fang Ku Shih]* (Traditions of the Imperial Workshops) has recorded the method of construction.

An interesting point here is that the invention is clearly attributed to the Early Han period rather than the Later. The book referred to is listed in the Hou Han bibliography, but perished, alas, long ago.

'Drum-carriages' were also known in the Han and perhaps earlier, though they are not positively called 'kilometre-measuring drum carriages' until the San Guo. It is probable, therefore, that such carriages in the Early Han were at first simply musical, intended for bands of musicians and drummers in state processions. The earliest seems to be that mentioned in the biography of Dan [Tan], prince of Yan Ci [Yen Tzhu], where we are only told that flags and banners were carried before and behind it; this would be about 110 BC. About 80 BC an important official, Han Yanshou [Han Yen-Shou], also had a drum-chariot in his processions. So did the Shanyu of the Southern Huns, and in AD 37 horses of particularly fine quality presented by foreigners were harnessed to the Han emperor's drum-chariot. The chapters on the imperial fleet of vehicles in the dynastic histories naturally associate the drum-chariot with people such as eunuchs, palace officials, attendants and familiars, actors, acrobats, conjurers, etc. While at first sight this would seem to strengthen the view that the early Han drum-chariots were purely musical, in fact it almost does the opposite, for we have already seen the intimate connection in ancient times between such entertainers and the makers of mechanical toys (volume 1, pp. 69–71 of this abridgement). The most reasonable supposition is that the drum-chariot was indeed originally a vehicle for musicians, but that some time in the Early Han (first century BC) the beating of drums and gongs was arranged to work automatically off the road-wheels, and that only then were the possibilities of such an instrument for surveying and charting itineraries realised. One pictorial representation has survived from the Han, namely that in the Xiaotang Shan [Hsiao Thang Shan] tomb series, dating from about AD 125. Carriages for musicians, whether mechanised or not, survived in imperial processions through many subsequent dynasties.

The question of the origin of the hodometer is important because parallel developments were occurring in Europe, as we shall see. The only extant Chinese specification, prepared by Lu Daolong [Lu Tao-Lung], Chief Chamberlain in the fifth year of the Tiansheng [Thien-Shêng] reign period (AD 1027), is in the *Song Shi*:

The Hodometer

... 'The vehicle should have a single pole and two wheels. On the body are two storeys, each containing a carved wooden figure holding a drumstick. The road wheels are each 1.8 metres in diameter, and 5.4 metres in circumference, one revolution covering 3 paces. According to ancient standards the pace was equal to 1.8 metres and 300 paces to a *li*; but now the *li* is reckoned as 360 paces of 1.5 metres each.

A vertical wheel is attached to the left road-wheel; it has a diameter of 42 cm with a circumference of 126 cm, and has 18 cogs 7 cm apart.

There is also a lower horizontal wheel of diameter 126 cm and circumference of 378 cm, with 54 cogs, the same distance apart as those on the vertical wheel (this engages with the former).

Upon a vertical shaft turning with this wheel, there is fixed a bronze "turning-like-the-wind" wheel which has (only) 3 cogs, the distance between these being 3 cm. (this turns the following one).

In the middle is a horizontal wheel, 1.2 metres in diameter, and 3.6 metres in circumference, with 100 cogs, the distance between these cogs being the same as the "turning-like-the-wind" wheel.

Next there is fixed (on the same shaft) a small horizontal wheel, 8.4 cm in diameter and 25 cm in circumference, having 10 cogs 3.8 cm apart.

(Engaging with this) there is an upper horizontal wheel having a diameter of 1 metre and a circumference of 3 metres, with 100 cogs, the same distance apart as those of the small horizontal wheel.

When the middle horizontal wheel has made 1 revolution, the carriage will have gone 1 *li* and the wooden figure in the lower storey will strike the drum. When the upper horizontal wheel has made 1 revolution, the carriage will have gone 10 *li* and the figure in the upper storey will strike the bell. The number of wheels used, great and small, is 8 in all, with a total of 285 teeth.

Thus the motion is transmitted as if by the links of a chain, the "dog-teeth" mutually engaging with one another, so that by due revolution everything comes back to its original starting-point.'

It was ordered that this specification should be handed down (to the appropriate officials) so that the machine might be made.

This description brings out clearly enough the reduction train of gearing and omits only the pegs on the shafts which tripped the wires to operate the puppets. Part of a model of such a hodometer is shown in Fig. 346. The

Fig. 346. Working model by Wang Zhenduo of a Han hodometer.

canopy of the carriage looks as if it may have been intended to revolve while the vehicle was in motion, and it would have been easy enough to make it do so.

A moment's attention to the last paragraph of Lu Daolong's report is worth while. It seems almost the fragment of a poem or essay, but the reference to 'dog-teeth' cogs in AD 1027 is important, since it shows that the Song engineers were conscious of the necessity of rounding them off, empirically foreshadowing the kind of shapes used today in mathematically defined involute and epicycloid gear teeth. Historians of engineering in Europe do not seem to know of any attention paid to cog shape before the time of Leonardo da Vinci (*c.* AD 1490). The oldest manuscript representation of gear teeth is about AD 1335, in the treatise of Guido da Vigevano, where the teeth are shown rounded, but this is later than those in Arabic and Chinese books.

The classical description of a hodometer in Europe was that of Heron of Alexandria (*c.* AD 60) though he did not claim it as a new invention; it had also been described by Vitruvius (*c.* 30 BC). The more complex of Heron's models recorded the distance by dropping balls into receptacles, and was used in the time of the emperor Commodus (AD 192), but after that there follows a very long gap, the next appearance being at the end of the fifteenth century AD in Western Europe. The pattern is therefore the same as that we have repeatedly met with, i.e. Greek antecedents, paralleled or followed at a short distance by Chinese developments which continue throughout the medieval period, and then a reawakening of the subject in Europe.

The hodometer, besides its method of construction which included a gear-train, used an auditory signalling system of jack figures beating on drums or gongs, and so was undoubtedly one of the precursors of all the jack-work of mechanical clocks.

THE SOUTH-POINTING CARRIAGE

If hodometers were widely spread, another kind of vehicle with gearing was peculiar to the Chinese culture-area. Allusion has already been made to the 'south-pointing carriage' in the section on magnetism (volume 3 of this abridgement), since it was long confused by Chinese and Westerners, with the magnetic compass. We know now, however, that it had nothing to do with magnetism, but was a two-wheeled cart with a train of gears so arranged as to keep a figure pointing due south, no matter what excursions the horse-drawn vehicle made from this direction. Yet it is none the less interesting for being mechanical and not magnetic; and if it was probably of little practical use (though we might be unwise to exclude attempted applications in surveying, as for its companion vehicle the hodometer), it has not deserved the summary liquidation which some modern scholars adept in other fields have given it.

The most important passage concerning the history of the south-pointing carriage is that in the *Song Shu* [*Sung Shu*] (History of the Song Dynasty) written *c.* AD 500.

> The south-pointing carriage was first constructed by the Duke of Zhou [Chou] (beginning of the first millennium BC) as a means of conducting homewards certain envoys who had arrived from a great distance beyond the frontiers. The country to be traversed was a boundless plain, in which people lost their bearings as to east and west, so (the Duke) caused this vehicle to be made in order that the ambassadors should be able to distinguish north and south.
>
> The *Gui Gu Zi* [*Kuei Ku Tzu*] (Book of the Devil Valley Master) says that the people of the State of Zheng [Cheng], when collecting jade, always carried with them a 'south-pointer', and by means of this were never in doubt (as to their position).
>
> During the Qin and Former Han dynasties, however, nothing more was heard of the vehicle. In the Later Han period, Zhang Heng [Chang Hêng] re-invented it, but owing to the confusion and turmoil at the close of the dynasty it was not preserved.
>
> In the State of Wei (in the San Guo period) Gaotang Long [Kaothang Lung) and Qin Lang [Chhin Lang] were both famous scholars; they disputed about the south-pointing carriage before the court, saying that there was no such thing, and that the story was nonsense. But during the Qinglong [Chhing-Lung] reign-period (AD 233 to 237) the emperor Mingdi [Ming Ti] commissioned the scholar Ma Jun [Ma Chün] to construct one, and he duly succeeded. This again was lost during the troubles attending the establishment of the Jin dynasty.
>
> Later on, Shi Hu [Shih Hu] (emperor of the Hunnish Later Zhao [Chao] dynasty) had one made by Xie Fei [Hsieh Fei]; and again Linghu Sheng made one for Yao Xing [Yao Hsing] (emperor of the (Jiang) Later Qin dynasty). The latter was obtained by emperor Andi [An Ti] of the Jin in the thirteenth year of the Yixi [I-Hsi] reign-period (AD 417), and it finally came into the hand of emperor Wudi [Wu Ti] of the (Liu) Song dynasty . . . Its appearance and construction was like that of a drum-carriage (hodometer). A wooden figure of a man was placed at the top, with its arm raised and pointing to the south (and the mechanism was arranged in such a way that) although the carriage turned round and round, the pointer-arm still indicated the south. In state processions, the south-pointing carriage led the way, accompanied by the imperial bodyguard.

These vehicles, constructed as they had been by barbarian workmen, did not function particularly well. Though called south-pointing carriages, they very often did not point true, and had to negotiate curves step by step, with the help of someone inside to adjust the machinery.

That ingenious man from Fanyang, Zu Chongzhi [Tsu Chhung-Chih], frequently said, therefore, that a new (and properly automatic) south-pointing carriage ought to be constructed. So towards the close of the Shengming [Shêng-Ming] reign-period (AD 477 to 479) the emperor Shundi [Shun Ti], during the premiership of the Prince of Qi [Chhi], commissioned (Zu) to make one, and when it was completed it was tested by Wang Sengqian [Wang Seng-Chhien], military governor of Danyang [Tanyang], and Liu Xiu [Liu Hsiu], president of the Board of Censors. The workmanship was excellent, and although the carriage was twisted and turned in a hundred directions, the hand never failed to point to the south . . .

The pigtailed barbarian Tuoba Dao [Thopa Tao] (third emperor of the Northern Wei dynasty) caused a south-pointing carriage to be constructed by an artificer called Guo Shanming [Kuo Shan-Ming], but after a year it was still not finished. (At the same time) there was a man from Fufeng [Fu-fêng], Ma Yo, who succeeded in making one, but when it was ready he was poisoned by Guo Shanming.

The first thing needing discussion here is the legendary material. The apparatus came in the course of time to be associated with two fabled events: (a) the battle between Huangdi and the great rebel Chi You [Chhih-Yu] when the latter made a smokescreen of fog through which the imperial army had to find its way, and (b) the sending home by Zhou Gong [Chou Kung] of the ambassadors of the Yueshang [Yueh-Shang] people to some place in the far south, for which guidance was necessary. It was thought that Han texts contained these stories, but though both events are described in two Han works, there is no mention of the south-pointing carriage. There is, however, one exception. A metaphorical reference in Liu Xiang's [Liu Hsiang's] *Hong Fan Wu Xing Zhuan* [*Hung Fan Wu Hsing Chuan*] (Discourse on the Hong Fan Chapter of the *Historical Classic* in Relation to the Five Elements) of about 10 BC. But Dr Needham suspects that this is one of those cases where the word for carriage was slipped in subsequently, the scribe not understanding that the 'south-pointer' reference had been to the lodestone spoon (volume 3, pp. 18 ff of this abridgement). The two legends, however, appear in developed form in the *Gu Jin Zhu* [*Ku Chin Chu*] (Commentary on Things Old and

New) by Cui Bao [Tshui Pao] about AD 300. This suggests that the south-pointing carriage was really a Later Han or Jin invention, and that the first machine was made, if not by Zhang Heng [Chang Hêng] about AD 120, then by Ma Jun [Ma Chün] about AD 255. There is no need to doubt the story about Ma Jun; he was an excellent engineer, and we are fortunate to have much information about him. We shall be meeting him again.

The making of the south-pointing carriage is recorded in other contemporary sources. Guo Yuansheng [Kuo Yüan-Shêng], in the *Shu Zheng Ji* [*Shu Chêng Chi*] (Records of Military Expeditions), a Jin book, says that outside the south gate of the capital there were Government Workshops and that the south-pointing carriage was normally garaged in the north gateway of this factory. He may well have been referring to Ma Jun's machine. It is interesting that Cui Bao's account ends by saying that the construction is described in a book now long lost, the *Shang Fang Gu Shi* [*Shang Fang Ku Shih*] (Traditions of the Imperial Workshops).

In the seventh century AD the invention reached Japan, as we hear of two monks constructing vehicles for the Japanese emperor in 658 and 666. In all probability these ecclesiastical engineers had themselves come from China. The south-pointing carriage was also combined sometimes with the hodometer.

The *Song Shi* [*Sung Shih*] (History of the Song Dynasty) of about AD 1345 gives the only detailed description of the machinery involved, as constructed by two engineers, Yan Su [Yen Su] in AD 1027, and Wu Deren [Wu Tê-Jen] in 1107. This text explains that besides two road wheels, the vehicle has seven gears, together having a total of 120 teeth. The largest gear wheel, almost 1.5 metres in diameter and possessing 48 teeth, was horizontal, and bore a vertical shaft nearly 2.4 metres high carrying a wooden figure pointing in a specific direction. After a full description of the gearing the text continues:

> When the carriage moves (southward) let the wooden figure point south. When it turns (and goes) eastwards, the (back end of the) pole (of the carriage) is pushed to the right . . .

At this point the internal gears perform part rotations causing the large horizontal wheel to rotate one quarter turn. This causes the pointing figure to indicate south once more. If the carriage be turned westwards instead, then the pole of the vehicle is pushed in the opposite direction – to the left instead of the right – so that the pointing figure is turned in the opposite direction to before, and so still ends up pointing south. Then the document continues:

> In the first year of the Daguan reign period (AD 1107), the Chamberlain Wu Deren [Wu Tê-Jen] presented specifications of the

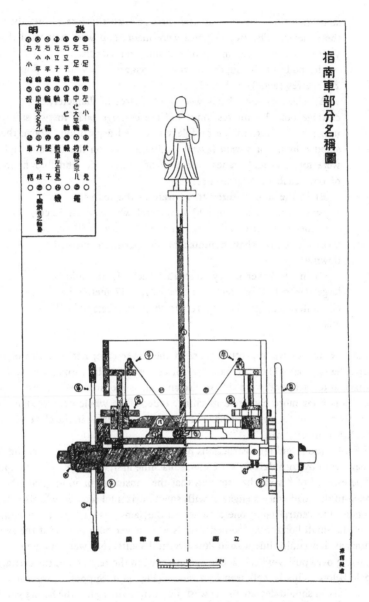

Fig. 347. Reconstruction of the mechanism of the south-pointing carriage according to A. C. Moule and Wang Zhenduo; back elevation. As the rear end of the vehicle's pole 8 moves to the left or right, it engages or disengages the suspended gear-wheels to 'the left and right respectively, thus connecting or disconnecting each road wheel with the central gear-wheel carrying the pointing figure.

south-pointing carriage and the carriage with the *li*-recording drum (hodometer). The two vehicles were made, and were first used that year at the great ceremony of the ancestral sacrifice.

The body of the south-pointing carriage was 3.9 metres (long), 2.9 metres (wide), and 3.32 metres deep.

(A) The carriage wheels were 1.74 metres in diameter, the carriage pole 3.2 metres long, and the carriage body in two storeys, upper and lower. In the middle was placed a partition. Above there stood a figure of a *xian* [*hsien*] holding a rod, on the left and right were tortoises and cranes, one each on either side, and four figures of boys each holding a tassel.

(B) In the upper storey there were at the four corners trip-mechanisms, and also 13 horizontal wheels, each 56 cm, 1.69 metres in circumference, with 32 teeth at intervals of 4.57 cm apart. A central shaft mounted on the partition, pierced downwards.

(C) In the lower storey were 13 wheels. In the middle was the largest wheel, 1.16 metres in diameter, 3.47 metres in circumference, and having 100 teeth at intervals of 3.17 cm apart . . .

There were pulleys and other gears in the lower compartment, two of the latter being small (33 cm in diameter), placed horizontally with an iron weight attached to each, and were able to rise and fall as required. As the text shows this was a far more elaborate machine. Indeed, the whole document is distinctly precious for our knowledge of Chinese engineering in the eleventh and twelfth centuries AD.

Explanations and reconstructions have been made in modern times, and we shall begin by considering Wu Deren's machine first, (see Figs. 347 and 348). The essence of his mechanism was that the subsidiary or inner gear-wheels fixed to the road-wheels engaged with small horizontal toothed wheels which rotated the central large one attached to the same shaft as the main figure. But the small horizontal toothed wheels were never both in gear at the same time; as they could slide up and down vertical shafts, they were hung on cords passing over pulley-wheels above and attached to the rear end of the carriage-pole below. This is well seen in a model (Fig. 349). Supposing the carriage going south should deviate to the west (i.e. turn to the right), the horses would carry round the pole to the right and hence its rear end would move to the left. This would raise the right small gear-wheel and lower the left one so that it enmeshed with the inner gear of the road-wheel and with the central wheel. As the left road-wheel would be moving while the right one would be out of gear, this train would obviously have the effect of compensating

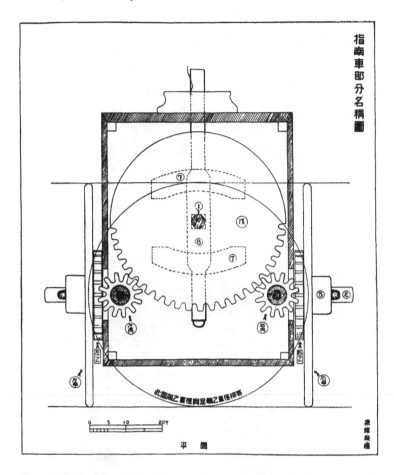

Fig. 348. Plan of the Moule–Wang reconstruction. The vehicle's pole rotates about the lower bearing of the vertical shaft carrying gear-wheel and pointing figure.

for the change in direction of the carriage and so maintaining approximately the south-pointing direction of the wooden figure. Upon resumption of a straight course, both gear-wheels would then be disengaged.

Yet Wu Deren's machine was in fact more complicated than this, since he introduced additional components into the gear-train, including (a feature of much interest for the history of technology) two-tier or double-gear wheels. Though why he multiplied the cogwheels is not very clear. Moreover, the reconstructions made in the light of the text may well not apply to the mechanisms of Yan Su and Ma Jun. And theoretically feasible though they

Fig. 349. Wang Zhenduo's working model with the large gear-wheel removed to show the position of the vehicle's pole and the cables which connect its rear end with the left and right rising and falling gear wheels. These slide freely on their vertical shafts, above which the small pulleys for the cables can be seen.

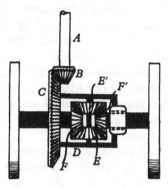

Fig. 350. Diagram to illustrate the principle of the differential drive of motor vehicles.

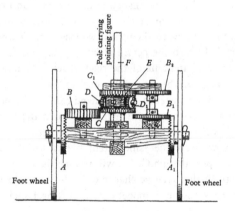

Fig. 351. Reconstruction of the mechanism of the south-pointing carriage according to George Lanchester. This back view shows that both road wheels are permanently connected with the pointing figure, which rides on a pair of idling wheels within the differential gear.

are, mechanically they are very inelegant, and the amount of play likely to have been involved, together with the difficulty of inducing the wheels to engage and disengage properly, would have made it hardly workable. An entirely different solution was therefore suggested by George Lanchester, himself an eminent engineer, who proposed that the machine (in some cases at any rate) embodied a simple form of differential gear.

The differential gear is an important element in the modern automobile. As has already become evident, when any wheeled vehicle rounds a corner, the outer wheels travel much further than the inner. The fixed, non-rotating, axles of carts or wagons permit each wheel to revolve independently, but on an axle through which power is transmitted, some device which allows the wheels to move independently while at the same time allowing force to be transmitted is essential. The differential gear of a power driven vehicle is shown in Fig. 350, though a true differential train, but not using bevel gears, was invented by Joseph Williamson about AD 1720 for adding or subtracting rotations to the mechanism of a mechanical clock. With the south-pointing carriage, the question now opened is whether a simple form of differential gearing goes back more than a thousand years earlier.

This is what Lanchester suggested, and it can readily be understood from Fig. 351 and the photograph of a model, Fig. 352. What is required is to reverse the direction of the drive to the vertical axle or post by the forces on the wheels in contact with the road. It is, of course, opposite to what occurs in an automobile. In Lanchester's reconstruction, the gears A, A_1 fixed to the road wheels engage with horizontal intermediate wheels B, B_1 (extended to B_2) but these are now arranged to be permanently in gear with an upper and lower bevelled wheel C_1, C; the upper one C_1 is concentric with the vertical shaft but not fixed to it. Between them run two small idling wheels D, D_1 connected by a stub-shaft from the centre of which rises the vertical shaft carrying the figure. Relative motions of the road-wheels will now be accurately, and inversely, reflected in the movements of the idling wheels and the pointer which they carry. If, for instance, the left-hand road wheel moves faster than the right, as on a westward turn, B will move while B_1, B_2 will be almost stationary, and C will be turning while C_1 will be almost at rest. Hence D, D_1 will react to precisely the same extent but in the opposite direction. They will then stay put, though C, C_1, will resume their mutually opposite regular motion until further relative change of road-wheel velocities occurs. Fig. 352 shows the behaviour of a model constructed by Lanchester himself.

Does this satisfy Yan Su's specification? If his two small wheels had teeth, it could. But a misprint in the text would have to be assumed, for he should have been recorded as speaking of two large horizontal wheels in the middle, not one. This could be due to a textual error easily made. There should also, however, have been ten or eleven wheels, and more than 120 teeth. Besides, if Yan Su's machine had used the differential gear, it is not obvious why the motions of the carriage-pole should have been mentioned at all, though the text gives no reason for bringing them in, and no cords or weights are mentioned. In any case, whether or not Yan Su used a form of differential gear, it remains quite possible that Ma Jun or Zu Chongzhi did so, for it is after all simpler as well as more practical and elegant than the clumsy device of engaging and disengaging gear-wheels vertically suspended.

Sinologists have generally regarded the south-pointing carriage as a kind of

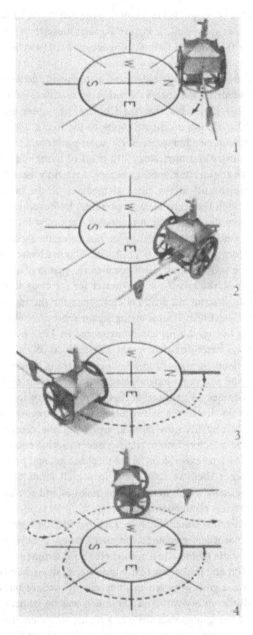

Fig. 352. Action photographs showing the efficiency of the mechanism of the differential-gear reconstruction of the south-pointing carriage. Pictures supplied by Mr Lanchester.

playful freak, or an exercise of misguided ingenuity on the part of the Chinese. But from a broad historical viewpoint, it is surely much more than this. It may be said to have been the first step in human history towards a cybernetic machine.

'Cybernetics', derived from the Greek *kubernetes*, a steersman, denotes the field of control and communication theory, whether in machines or living organisms. Earlier (pp. 101 ff) we saw that the Chinese, like other ancient peoples, were intrigued by making inanimate copies of the living body, and constructed many automata depending on springs, water-power and the like. Though classical biochemistry and physiology still think of living organisms in terms of power and fuel engineering, modern biology has firmly established the existence of a 'homoeostasis' of the internal medium of the body – a remarkable constancy of such factors as osmotic pressure, hydrogen ion concentration, blood-sugar, etc. So also modern technology has now for a long time past developed all kinds of self-regulating or homoeostatic machines.

Its first truly practical form emerged in AD 1712 with Thomas Newcomen's steam engine in which the reciprocating beam actuated the piston of a water-pump which on its down-stroke raised the cold water for injecting into the cylinder. From this point onwards the road lay clear open for the invention of the slide-valve for single-cylinder double-acting steam engines by William Murdock (1799). This device used a principle introduced in 1782 by James Watt, which itself closely resembled that in the characteristically Chinese ancient single-cylinder double-acting piston-bellows – with of course the essential difference that the piston no longer acted as a compressor on both strokes but was itself subjected to pressure on both strokes. Here then was a self-acting cyclical system, but not a self-regulating one. However, Watt accomplished this in 1787 by his now familiar device of the centrifugal governor for regulating the velocity of steam engines under varying conditions of load. Thus we arrive at a true closed-loop 'servo-mechanism', and probably the oldest of its kind. Since then we have come to use all kinds of other 'feedback' devices – thermostats, computers, homing missiles and self-adjusting chemical and engineering production equipment.

This is the background against which it may be said that the south-pointing carriage was the first homoeostatic machine in human history, involving full negative feedback. With it the Chinese had constructed an instrument which would fully compensate for, and thus consistently indicate, all deviations from a prescribed course. It was a great achievement, not only of practice but also of conception, for the third century AD. And perhaps it was no coincidence that this simulacrum of a living organism arose in a highly stable civilisation characterised by a highly organic conception of nature, for the tendency to self-regulation is a primary property of living organisms.

ANIMAL TRACTION AS A POWER SOURCE

So far we have been concerned with the use of animal-power as the prime mover for various kinds of machines, as also for vehicles, and it was convenient to leave vague the exact practical way in which this force was applied. Those who were really concerned with the matter, however, in different historical periods, found themselves face to face with a set of problems which were really of an engineering character. Though we do not usually think of it in this way, any system of harnessing constitutes an elaborate play of linkwork and hinges, in which both the anatomical nature of the draught animal and the structure of the object pulled has to be taken into consideration.

Efficient harness and its history

It was the French cavalry officer Lefebvre des Noëttes who in 1931 pointed out that a draught system composed of one or more animals, should have a rational method of harnessing. This should permit the complete use of the force of all the animals, and favour their work as a team. This is so in the 'modern' or collar-harness, which is well adapted to the anatomy of the horse. The 'antique' harness, on the other hand (which we shall call the 'throat-and-girth' harness), could make use only of a small fraction of the possible motive force of each animal and failed to ensure satisfactory collective effort. In general its efficiency was four to five times less than that of the breast-strap, and it finally died out in Western Europe in the Middle Ages. Moreover, it was the same everywhere, in all ancient realms and cultures, equally inefficient. Only one civilisation broke away from this and developed an efficient harness – China.

The throat-and-girth harness is shown in Fig. 353(*a*). It consists of a girth surrounding the belly and rear part of the rib region, at the top of which the point of traction is located. Presumably in order to prevent the girth from being carried backwards, the ancients combined it with a throat-strap, sometimes narrow, more often broad, which crossed diagonally the ridge between the horse's shoulder bones (the 'withers'). The throat-strap so compressed the animal's throat that it suffocated the horse as soon as it attempted to put forth a full tractive effort. In sharp contrast to this is the collar-harness (Fig. 353(*c*)). Here the collar is reinforced and padded, bears directly upon the breast and the muscles which cover it, thus linking the line of traction intimately with the skeleton of the horse and freeing fully the throat for breathing. The animal is now able to exert its maximum tractive force,

The collar-harness was not, however, the only way in which the line of traction could be derived from the chest so as to leave the throat free. At some time during the Qin or early Han in China, or quite probably earlier, in the Warring States period, someone realised that the animal's shoulders could be surrounded by a strap, which, if suspended by a strap across the withers, and

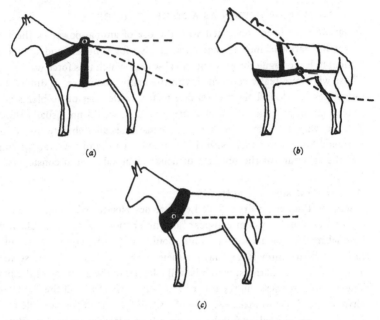

(a)

(b)

(c)

Fig. 353. The three main forms of horse or mule harness. (a) The throat-and-girth harness, characteristic of Western antiquity. Compression of the windpipe prevented efficient pulling power. (b) Breast-strap harness, characteristic of ancient and early medieval China. By pressure upon the breast the line of traction is directly linked with the animal's skeleton, and full pulling power could be exerted. (c) Collar harness, used from later medieval times both in China and the West. As the breast region is again the bearing point, efficient pulling power can be exerted.

attached also to the point where the bend of the curving cart or chariot shafts changed direction, would greatly increase the efficiency of the horse's work. This is the breast-strap harness shown in Fig. 353(b). The continuation of the harness round the animal's hind-quarters, and its support by a hip-strap, was not a necessary part of the traction system, but permitted a backward movement of the cart, and its braking when descending slopes; to this point we shall return.

Throat-and-girth harness in Sumer and Shang
One might say that the throat-and-girth harness of horses was a makeshift alternative for the yoke of the ox. Since the ox's neck projects forwards from the body horizontally, unlike the rising crest of the horse, and since its spinal column forms a bony contour in front of which a yoke can easily be placed, it was satisfactory enough in the earliest times. Fig. 354 shows an ox-cart of

Fig. 354. The age-old ox-yoke harness for bovine animals, which have a kind of hump against which the arched yoke can bear. No such ledge is available in horses. Photographed in 1958.

age-old type in the streets of Jiuquan [Chiu-chhüan] (Gansu [Kansu] province) in 1958. Abundant illustrations of similar vehicles drawn by oxen are to be found in Han reliefs, Buddhist carvings and Dunhuang [Tunhuang] frescoes. But the yoke on each side of a central pole, or between shafts, which was suitable for oxen was quite inapplicable to horses, and the throat-and-girth harness was the consequent substitute attachment.

In the Shang period and no doubt for several centuries afterwards, the junction point of girth and throat-band was bound by leather thongs to the cross-bar attached to the middle of the upward-curving central pole, transmitting thus the tractive force. But for some reason or other, perhaps of ornament, a trace of the useless 'yoke' lived on with great persistence. Before the beginning of the Zhou it had assumed a narrow V-shaped 'wishbone' form, and its lower ends were turned up, perhaps to act as guides for the reins. Then at some time about the Warring States period the Chinese abandoned the central pole in favour of two parallel outside bars, and the curving S-shaped shafts of the typical Han chariot were the result. Still, however, the 'fork-yoke' persisted, probably because it was still useful as a guide for the reins. This

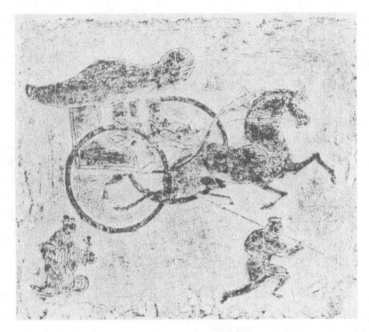

Fig. 355. A form of yoke still persisting in Han times. It is fixed at the centre of the cross-bar connecting the shafts of a light luggage trap, and its ends seem to guide the reins. From a moulded brick and now in the Chongqing Municipal Museum.

appears to be the arrangement in Fig. 355 taken from a moulded Han brick found in Sichuan.

After the invention of the breast-strap harness, and its attachment to the point on the shafts where the curve changed direction, the cross-bar was of very little use, and eventually it disappeared. So too did the 'fork-yoke'. Here we meet with a question of considerable interest, namely the dates of the two inventions – the shafted vehicle and the breast-strap harness. Did they coincide, or if not what interval separated them, and which came first?

The really astonishing thing about the throat-and-girth harness is the immense spread which it had in both space and time. We find it first in the oldest Chaldean representations from the beginning of the third millennium BC onwards, in Sumer and Assyria (1400 to 800 BC). It was in sole use in Egypt from at least 1500 BC, where it is shown on all paintings and carvings of chariots and horses, and it was likewise universal in Minoan and Greek times. Innumerable examples occur in Roman representations of all periods, and the empire of the throat-and-girth system also covered Etruscan, Persian

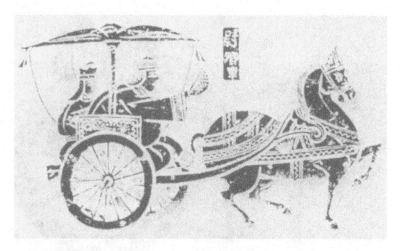

Fig. 356. Han breast-strap harness, a typical representation; the carriage of the Intendant of the Imperial Equipages, from the Wuliang tomb-shrines (*c*. AD 147). Besides the breast-band or trace supported by a withers-strap (here apparently forked) and hip-strap, a girth is still retained and the old throat-strap has turned into an elongated sling helping to suspend the front of the chest-strap. Many Han representations are less elaborate than this, showing only the chest-strap with its suspensions.

and early Byzantine vehicles without exception. Western Europe knew nothing else until about AD 600, nor did Islam. Moreover, the south of Asia was almost entirely in reliance on this inefficient harness, for it is seen in most of the pictures of carts which we have from ancient and medieval India, Java, Burma, Siam and other parts of that area. Central Asia, too, has it. One of its last appearances occurs on a bas-relief of the fourteenth century AD at Florence in Italy, where it may be a conscious reference to the past.

The first step: breast-strap harness in Chu and Han

We may now pass to the efficient breast-strap harness of China with its two straps or 'traces' across the breast suspended by the withers-strap (Fig. 353(*b*)). One can only say that it is universal on all Han carvings, reliefs and stamped bricks which show horses and chariots. A splendid Han bronze model of a harnessed horse and cart (Fig. 357) shows clearly the breast-strap and the shafts, while a relief from the Yinan [I-nan] tomb-shrines (Fig. 358) shows many details, which are elucidated in Fig. 359. In this carriage from the end of the second century AD, the shafts of the vehicle are now split in two, for at or near to the point of attachment of the traces a short rod is bound to them. This keeps the lower ends of the 'fork-yoke', which is attached centrally

Fig. 357. Bronze model of a light luggage trap like that of Fig. 355 but without the roof, Han period. Though there is neither breeching nor other parts of the harness, the breast-strap attached to the mid-point of the curving shafts is well seen. In the Hosokawa Collection, Tokyo.

Fig. 358. Another carriage harness relief of the Han period, from the Yinan tomb-shrines, *c.* AD 193. Compare with Fig. 359.

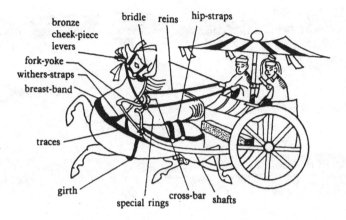

Fig. 359. Elucidation of the Yinan relief to show the various components.

to the cross-bar, pointing forwards and downwards. The reins run through rings on the cross-bar, and their front ends are attached to bronze cheek-piece levers at each end of the bit. The bridle, or network of straps round the horse's head, is also clearly shown by the carvers of Yinan. The traces themselves were generally attached to the mid-points of the shafts, but it is clear from some of the reliefs that in the case of heavy baggage-carts and similar vehicles, they were run right back to the body of the vehicle itself.

An interesting piece of evidence concerning the origins of the breast-strap harness is to be found on a painted lacquer box from the State of Chu, of approximately fourth century BC. It seems to show a transition between the throat-and-girth harness and the use of straps (Fig. 360). But the throat-strap has not been quite abandoned, only lengthened so as to support the chest piece (or pieces) at the change of curvature of the shafts. This acts like a withers-strap which became used to support the traces. The girth-strap has also been retained (Fig. 361). This evidence would perhaps suggest that the invention of the shafted chariot was the limiting factor, and that traction from the breast-bone area was not achieved until the shafts became available.

What gave the idea for the breast-strap harness we do not know, but there is always the possibility that it derived from what was found most convenient in human haulage. The working of boats upstream by large groups of trackers goes back a long way, no doubt, in Chinese history. Men would have been conscious from their own experience that a tractive force must be exerted from the back and collarbone regions of the body in such a way as to permit breathing freely. Tracking is represented in the Dunhuang [Tunhuang] frescoes, from which Fig. 362 may enhance this argument.

Fig. 360. A horse and chariot from a round lacquer box of the kind made in the State of Qu, *c*. fourth century BC. A hard yoke-shaped object surrounds the horse's chest and is connected to the mid-point of the shafts by traces. This is a possible intermediate form between the throat-and-girth harness and the breast-strap harness. Found in a Han tomb near Changsha.

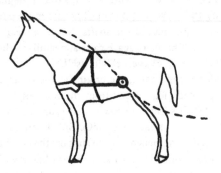

Fig. 361. Diagram to explain the harness seen in the Changsha lacquer painting.

Comparative estimates

It is hardly necessary to insist on the advantages of the horse, once properly harnessed, over the ox. While both animals have about the same pulling power, the horse gives some 50 per cent more kilogramme force per second because it moves more quickly. Besides this, the horse has greater endurance than the ox, and can work a few more hours each day.

Experiments made by des Noëttes in 1910 showed that the throat-and-girth harness limited the load two horses harnessed together could pull to about

Fig. 362. Trackers hauling a boat upstream; fresco painting of the early Tang period from cave no. 322 at Qianfodong, Dunhuang. Efficient horse harness may have arisen from the practices of human hauliers.

half a tonne, though with a collar harness (which we shall discuss in a moment) this rose to nine tonnes. Indeed, horses with the throat-and-girth harness could not pull many modern horse-drawn vehicles even if unloaded. But the breast-strap harness made a considerable difference. Moreover, an attentive study of Han illustrations of chariots and carriages, comparing them with other ancient civilisations, clearly shows that the Chinese vehicles were much heavier. While Egyptian, Greek or Roman chariots always appear of minimal size, fit only for two persons at most, with cut-away sides and often drawn by four horses, the Chinese chariots often show as many as six passengers. Very frequently too they have heavy upcurving roofs (see Fig. 355 p. 204) and are usually drawn by only one horse. Again, when the Chinese came into contact with Western regions, they distinctly noticed the vehicles of those parts as being remarkably small, and the chapter on Arabia and Roman Syria in the *Hou Han Shu* (History of the Later Han Dynasty) of AD 450 says just this.

What is more, in the *Mo Zi* [*Mo Tzu*] (The Book of Master Mo) of the late fourth century BC, we read that when Mo Di [Mo Ti] comments on a flying bird-like kite constructed by Gongshu Pan [Kungshu Phan] he says:

Your achievement in constructing this bird is not comparable with
that of a carpenter making a linch-pin (for a cart). In a few
moments he cuts out a piece of wood which, though only
7.5 cm long, can carry a load no less than 50 *dan* [*tan*] in weight.

In a parallel passage in *Han Fei Zi* [*Han Fei Tzu*] (The Book of Master Han
Fei), which would date from the following century, the figure given for the
weight of the cart pulled is 30 *dan*. Since the *dan* of the Warring States period
is estimated as having been equivalent to over 54 kilogrammes, Mo Di's
estimate would amount to some 2.7 tonnes, and that of Han Fei to rather
over 1.6 tonnes. Figures such as these certainly seem to support the view that
the breast-strap harness was little less efficient than the collar-harness itself.

That the throat-and-girth harness was often felt to be unsatisfactory in antiquity is shown by numerous unsuccessful attempts to improve it. In ancient
Egypt there was a strap which descended from the band round the throat,
passed between the horse's front legs and ended up attached to the girth. This
was very little use, because the pull still remained on the horse's back so that
the throat-band always pulled upwards. A second attempt was what may be
called a 'false breast-strap'. Here a horizontal band was placed round the
horse's chest, fixed to the girth at each end. This was not much use either,
since the pull caused it to rise and compete with the throat-strap in strangling
the horse. Yet it was tried again and again, from Assyria in the eight century
BC to Byzantium in the ninth century AD. A third method was connected
with the use of the saddle on draught horses. The throat-strap was attached
either further back towards the horse's loins being connected with the rear
of the saddle, or a little lower down but still attached to the rear side of the
saddle. This was in Byzantium from the tenth to thirteenth centuries AD and
at the other end of the Old World in Cambodia during the twelfth century
AD, and it found its way through to modern Japanese ploughs and North-
West Indian light two-wheeled carts or tongas. But still the problem remained
unsolved. Except in China.

Radiation of the inventions

In due course the Chinese breast-strap harness of the Han arrived in Europe.
So far as documentary evidence goes, there is no European representation of
this before the eight century AD. After this it appears more frequently, as for
instance in a ninth-century tapestry in the Norwegian Oseberg ship-burial,
among the Vikings of the ninth and tenth centuries, and again as an illustration in a manuscript dated AD 1023 of *De Rerum Naturis* (On the Natural
Qualities of Things) of the German Hrabanus Maurus (AD 776 to 856). It
seems also to be depicted on the Bayeux Tapestry (AD 1130) after which it
becomes widely used, not least on post-coaches of the early nineteenth
century.

It has long been known that the other efficient harness, the collar-harness, made its appearance in Western Europe about the beginning of the tenth century AD, and had been universally adopted by the end of the twelfth. Manuscript pictures of about AD 920 are the first to show it, and afterwards it becomes increasingly common. There is no doubt that over the same period it was largely replacing the old breast-strap harness in China. It is seen in Song and Yuan paintings (the eleventh and fourteenth centuries), and appears to the exclusion of all other types of harnessing in the collections on agricultural engineering already described (p. 107 above). But soon a suspicion arose that the collar-harness goes back a great deal further in China and Central Asia than in Europe.

It is now possible to attain greater precision concerning the time and place of the cardinal inventions and their subsequent arrival in Europe. Certainly

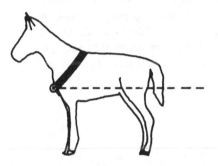

(a) The common element in the Chhien-fo-tung frescoes.

(b) Chhien-fo-tung frescoes; three-component representations.

(c) Contemporary collar-harness of north and north-west China.

(d) Chhien-fo-tung frescoes; two-component representations.

Fig. 363. Diagrams to explain the evidence of the fresco paintings at the Qianfodong cave temples, Dunhuang (fifth to eleventh centuries AD).

there can be no doubt that the earliest Chinese chariots were of the pole type, with throat-and-girth harness, as is shown by the analysis of the written character as well as the archaeological evidence. This system probably lasted as late as the end of the Warring States period, but was completely superseded by the breast-strap harness during the early Han (i.e. from the second century BC onwards). In no other part of the world did an efficient harness for horses appear so early. That the breast-strap harness then made its way in due course to the West across the Eurasian plains we can hardly doubt. There is in fact evidence that it reached Italy at least three centuries before the rest of Europe; possibly with the eastern Goths (Ostrogoths) in the early fifth century AD.

But could this be true of the second efficient horse harness also? That the collar was known in Tang times, that is two or three centuries before its first representation in Europe, is evident in the cave-temples near Dunhuang. Here in nearly all the representations of horses and carts the shafts are attached to the lowest and furthest forward point of the collar-like harness (Fig. 363(*a*)). This is radically different from what is seen in drawings and carvings from all other parts of the world, where the attachment, if not in the throat-and-girth position on the horse's back, is to the mid-point of the throat-strap or collar-like band. It bears the clear implication of a pull from the chest and not the region of the throat.

Fig. 364. Copy-painting of part of the painting of the procession of Zhang Yichao's wife, which shows an enlarged picture of a baggage cart. The curved 'ox-yoke' cross-bar connecting the shafts fits across the front of a well padded ring-shaped cushion passing low across the horse's chest. The collar, in fact, is substituting for the 'hump' of the ox.

The second step: collar harness in Shu and Wei

Let us begin with the clearest case. In cave no. 156, a magnificent panorama depicts the triumphal procession of the wife of a Chinese general and provincial governor Zhang Yichao [Chang I-Chhao] who recovered the Dunhuang region from the Tibetans in AD 834. There is evidence that this was painted in AD 851. Following the commander of the army is a retinue which contains four carts, three of which are for baggage. One of these is shown enlarged in Fig. 364. In all cases they have three components: (i) the shafts, (ii) a curved piece of wood like the ox-yoke or the cross-bar of the Han carriage which connects them together, and (iii) a well-padded collar coming low on the chest (Fig. 363(*b*)), and rising behind the cross-bar. It thus becomes immediately evident that in these earliest forms of collar-harness the arrangement was in two parts, and the collar alone was simply an artificial substitute conferring on the horse the internal 'hump' of the ox, thus giving a point on which the yoke could rest.

Having seen these paintings, Dr Needham was delighted to find along the roads of Gansu province and many other parts of North and North-West China that the form of collar-harness still widely in use there perpetuates the double system of the Tang. Fig. 365, a photograph taken in 1958 near Jiuquan [Chiu-chhüan], shows it well. The collar is not itself attached to anything, but here rests upon a clasping framework, the descendant of the ancient yoke, which is directly connected with the shafts by thongs slipped over the pegs which they carry at their front ends (Fig. 363(*c*)). In some districts the yokes of oxen assume the shape of a wishbone or a tree-branch forked in a wide angle. In this we may see a transition to the 'yoke' necessary for the horse in the collar-harness invention, and we shall naturally remember also the wishbone-shaped rudimentary 'yokes' of the horse-drawn chariots of the Han. Thus the essence of the new device was an 'artificial hump' which would not come off. Only later on were the hard yoke and soft collar combined into one component, as in a modern harness, to which the Chinese dictionary term for the horse-collar, *hujian* [*hu-chien*], or 'shoulder-protector', applies. And our original question can now be answered, in principle if not in date, for many of the earliest European pictures of collar-harness show precisely the Chinese three-component system. Indeed one can actually meet with it still in certain corners of Europe; in 1960 Dr Needham found it again in the Iberian peninsula at the furthest extreme of the Old World.

Many other pictures of horses and carts are seen in the caves near Dunhuang. All show attachment of the shafts to the furthest forward point of the object round the shoulders and chest. Some form of collar-harness was therefore clearly implied even without the crucial evidence of cave no. 156. There need be no hesitation in assuming the existence of the padded 'hump-substitute' when the drawings do not show it. Otherwise the 'horse-yoke' could not have worked at all. And indeed while the whole three-component system

Fig. 365. Typical collar-harness still in use in North and North-West China (1958). The collar, still called the cushion, sits free round the neck of the horse or mule, and takes the pressure of a clasping framework, the direct descendant of the ancient yoke, which is attached to the shaft by thongs slipped over pegs at their front end. In this photograph taken on the edge of the Gobi desert, two leading animals were adding their power by means of traces.

(Fig. 363(*b*)) is seen in at least four or five caves, there are at least five or six where we find simply the shafts and the yoke (Fig. 363(*d*)). Again, by good fortune, the oldest representation of all (cave no. 257), painted under the Northern Wei between AD 477 and 499, shows the collar-harness clearly (Fig. 366). Reins are indicated by a pair of lines, high up on the neck, or just possibly a collar not drawn in its proper place behind the yoke – yet the evidence is decisive, for the yoke itself is well placed, and though no collar appears for it to rest on, it would have been useless without it. The same argument applies to the next oldest pictures, precisely datable between AD 520 and 524. Then there are two further two-component frescoes both from the Sui dynasty (*c.* AD 600), which bring us back to the certainties of

Fig. 366. The oldest representation of the equine harness in the Qianfodong cave temple frescoes, painted under the Northern Wei between AD 477 and 479. A cart with a 'sun bonnet' roof of the covered-wagon type complete with curtains and a sun awning, is being drawn by a large horse. The arching cross-bar is clear, but the artist failed to draw the cushioning collar behind it, without which it would have been useless. Two thin lines higher up the horse's neck may be a collar drawn in the wrong place but are more probably reins, and the breech-strap is indicated by another.

the Tang. There are also certain scenes when animals other than horses are harnessed by yokes within shafts. One of these is of Tang date (probably early ninth century AD); here the artist may be illustrating a legendary story, but the mechanical principles remain in force. After this the representations, though generally clearer, parallel or follow the presumed earliest appearance of the collar-harness in Europe.

The conclusion from the Dunhuang evidence is thus quite clear. Collar-harness in its initial form appears explicitly and indubitably in the procession of Zhang Yichao in AD 851, nearly a century before the first possibly acceptable documents of Europe. But depictions which can only have been based upon this harness go back to the last quarter of the fifth century AD and the first quarter of the sixth, so that it would be very reasonable to date the first appearance of it about AD 475 in the empire of the Northern Wei. And the place would again be significant, for the sands of the borders of the Gobi desert in Gansu and Shanxi needed strong tractive apparatus. Where Han trace-harness would break, the clasping framework could be attached to the shafts by chains, and thus at last the sheer muscular strength available became the only limiting factor.

A third source of information is constituted by bricks from the tombs of the Qin, Han and San Guo periods. It would still be premature to fix upon the late fifth century AD as the assured time of origin of the collar-harness invention, for we may have to recognise it as the third century AD. During the First World War excavations of the tomb of Bao Sanniang [Pao San-Niang] at Zhaohua (Chaohua) on the Sichuan – Shenxi road, brought to light a series of magnificent solid moulded bricks, each of which has the same picture of a horse and cart (Fig. 367). As they are not all equally damaged or imperfect in the same places, it is possible to make out what seems to be a large horse-collar coming low on the chest, apparently amply padded and looking almost like a thick garland. The straight shaft or traces seem to be attached to the side of this with no sign of a yoke. In some cases it is also possible to see a faint girth and perhaps the horizontal line of the breeching – the strap round the horse's hindquarters which enables it to hold back the vehicle. Even the reins appear in some of the bricks. If collar-harness is really present the fact is indeed remarkable, for the date of the tomb is unquestionably in the San Guo period, between AD 221 and 265. But the bricks of Bao Sanniang do not stand quite alone, for in a rare collection of reliefs and and brick rubbings mostly of unknown provenance published during this century by Di Pingzi [Ti Phing-Tzu], there is one which shows a chariot of three horses which have thick and distinctly ring-like objects round their chests). Unfortunately, the origin of this hollow moulded brick is uncertain, but there is no reason for doubting its genuineness. It cannot be later in date than the Early Han, but could be as old as the Warring States, so we can only place it between the fourth and first centuries BC. Possibly, therefore, the occasional use of the collar-harness goes back in China far earlier than its general adoption in the sixth century AD. The position may be summarised in the following way:

Form of harness		Des Noëttes documentary representations)	Newer evidence	Haudricourt (philological evidence)
Throat-and-girth		Very ancient	Very ancient	–
First breast-strap	Ch.	1st AD	4th to 2nd BC	–
	Eu.	12th AD	8th AD	5th or 6th AD
First collar-harness	Ch.	17th AD	1st AD (Han bricks) BC	–
			3rd AD (San Kuo bricks)	
			5th AD (Tunhuang frescoes)	
	Eu.	10th AD	–	9th AD

Fig. 367. Some of the moulded bricks in the tomb at Zhaohua, Sichuan, dated between AD 220 and 260. They depict something very like the collar-harness. An amply padded object like a large garland surrounds the withers and passes across the back region at a low level. In one brick or another, undamaged in different places, shafts or traces and reins or breeching can be made out, together with a distinct girth. If a decorative garland is not simply obscuring normal breast-strap harness here, the invention of collar-harness may have been made in Sichuan.

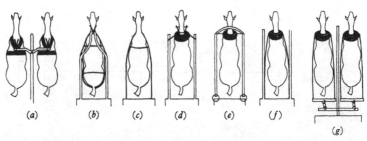

Fig. 368. Modes of attachment of equine harness to vehicles or machinery.
(*a*) Pole, cross bar, yokes and throat-and-girth harness; (*b*) Han breast-strap
harness with shafts, rudimentary yoke, hip-strap and breeching present;
(*c*) postillion or later breast-strap harness with traces attached to vehicle;
(*d*) traditional Chinese collar-harness with the hard component, descendant
of the yoke and ancestor of the hames, attached directly to the front ends
of the shafts; (*e*) *duga* of Russia and Finland, retaining the arched cross-bar
because the shafts are not structurally part of the vehicle; (*f*) modern
collar-harness, the traces attached directly to the vehicle; (*g*) whippletrees
for receiving the pull of the traces of collar harness.

It only remains to glance briefly at the relations of these different types of
harness to their vehicles. The throat-and-girth harness, with its pole, cross-bar
and fork-yokes seen from above, shown in Fig. 368(*a*), is to be contrasted with
Fig. 368(*b*) the Han system, where the fork-yoke had not disappeared but had
become even more unnecessary because of the full use of the breast-strap and
traces. It was the predecessor of the simple breast-strap ('postillion') harness
of later times (*c*). But after the Chinese adopted the collar-harness, one special
feature persisted until the present day, namely the attachment of the hard
component of the collar directly to the ends of the shafts (*d*). This, derived
from the ancient ox-yoke method, was only possible because in the Chinese
cart the shafts always formed a continuous part of the structure of the vehicle.
If this is not the case, it becomes necessary to keep the ends of the shafts
away from one another by a special bar, since they tend to be pulled together
in use. This is seen in the Russian or Finnish *duga* (*e*).

The problem is avoided by the attachment of the traces to the vehicle itself,
as in some ancient Chinese and in modern European usage (Fig. 368(*f*)). This
has the further advantage that more than one horse can be attached in file,
using the front end of the shafts. Meanwhile the hames or projections from
traces had fused with the collar (*f*) and (*g*). Then in the West the yoke, no
longer necessary at the level of the horse's back, moved downwards, either
in front of the chest or underneath the horse's belly, or behind the animal
altogether to become eventually a bar – the 'whippletree' (*g*). In the twelfth
century AD this was fixed firmly across the pole, but later it was linked to
two or more movable whippletrees, as also shown in (*g*).

Animal power and human labour

This chapter cannot end without a reference to the social aspects of efficient equine harness, and the use of animal power instead of human labour, which have been much debated. Though it has been proposed that the appearance of the collar-harness led to the decline and disappearance of slavery, it has also been pointed out that mass slavery vanished from Europe before this type of harness ever arrived. But this has never affected the converse argument that after all no efficient form of horse harness was invented in those ancient Mediterranean societies which embodied mass slavery and presumably abundant human labour. If the invention of an efficient harness was clearly not the cause of the decline of slavery, perhaps the presence of slavery inhibited the invention?

It certainly seems one of the paradoxes of history that in spite of the intellectual brilliance of the Greek philosophers and remarkable ingenuity of the Alexandrian mechanicians, the ancient Western world never succeeded in solving the problem of horses efficiently. Perhaps of course they did not try. However, it must be remembered that from the fifth century BC onwards Greek and Roman engineers moved all their heavy weights (involving loads of up to some eight tonnes) quite effectively by the aid of yoked oxen, often in file. It has therefore been suggested that Hellenic and Hellenistic civilisation did not in fact regard the horse as a draught-animal at all – it was primarily aristocratic, swift and military; and it had always been in short supply. Yet even so, one would have thought that for some purely military purposes an efficient equine harness could have been decidedly useful. In any case, for the Chinese and their Hunnish and Mongol neighbours, ever in the saddle, horses were assuredly not so rare and precious. This is a feature of northern Chinese culture which may well be relevant to the fact that both the efficient forms of harness for traction by horses were invented in this region.

Here a comparison of Hellenistic slave-owning society with ancient Chinese society inevitably presents itself, but the complex problems involved can only be alluded to here. It may well be that Chinese society incorporated only domestic slavery, and that the proportion of slaves to the total population (in the Han, for instance) was always small; but the available human-power liable to *corvée* or duty labour was, at any rate in some periods, abundant enough. Possibly the monsoon climate, with its consequence of strictly fixed seasonal agricultural work, from which Confucian morality and common sense alike forbade the abstraction of the toiling peasant-farmers, led to the use of greater ingenuity in solving the mechanical problems of efficient traction, which might arise at any time of year. In the last resort it was perhaps the nomadic peoples who first faced and solved some of them, aided and stimulated at the borders of culture-contact by the practical genius of the Chinese.

5

Clockwork: six hidden centuries

The clock is the earliest and most important of complex scientific machines. Its influence upon the world-outlook of developing modern science from Nicole Oresme in the fourteenth century, when he likened the heavens to a mechanical clock, was incalculable.

No one can doubt that the invention was one of the greatest achievements in the history of all science and technology. As the historian H. von Bertele wrote some forty years ago, 'The fundamental solution of the problem of securing steady motion by intersecting the progress (of a weight-driven or any powered train [of gears]) into intervals of equal duration, must be considered as the work of a brain of genius.'

The fundamental engineering task was to devise means of slowing down the rotation of a wheel so that it would keep a constant speed continuously in time with the apparent daily revolution of the heavens. The essential invention was the escapement. In what follows we shall show that the first of all escapements arose in China in the middle of a very long line of development of mechanisms for the slow rotation of astronomical models to demonstrate the rotation of the heavens, though their primary aim was for computation rather than time-keeping as such. We shall also show that its first application was to a water-wheel like that of a vertically mounted water-mill, so that although in later ages mechanical clocks were mainly driven by falling weights or expanding springs, their earliest representatives depended on water-power. The mechanical clock thus owes its existence largely to the art of Chinese millwrights. This differs widely from the account accepted hitherto. How is it that the Chinese contributions to clock-making have been hidden from world history?

It will readily be allowed that few historical events were so rich in consequence as the decision taken by certain southern Chinese officials in AD 1583 to invite into China some of the Jesuit missionaries who were waiting in Macao. It was the first decisive step in the long process of unification of world science in Eastern Asia, and the better mutual understanding of the great cultures of China and Europe. The two men chiefly concerned were Chen Rui [Chhen Jui] (1513 to *c.* 1585), who was for a short time Viceroy of the two Guang [Kuang] provinces, and Wang Pan [Wang Phan] (1539 to *c.* 1600) who was Governor of the city of Zhaoqing [Chao-chhing]. They were particularly interested in reports that the Jesuits had or knew how to make, chiming clocks of modern type, i.e. of metal with spring or weight drives and striking mechanisms. These became known as 'self-sounding bells', *zi ming zhong* [*tzu ming chung*] (自鳴鐘), by a direct translation of the word 'clock' or *cloche, glocke.* This is important, for an entirely new name, as this one was, naturally suggested an entirely new thing. The mechanical clocks of the Chinese Middle Ages had been, as we shall see, extremely cumbrous and probably never very widespread; moreover no special name had distinguished them from non-mechanised astronomical devices. It was therefore not surprising that the majority of Chinese, even scholars in official positions, now got the impression that the mechanical clock was a new invention of dazzling ingenuity which European intelligence alone could have brought into being. And of course the missionaries (as men of the Renaissance) quite sincerely believed in this higher European science, seeking by analogy to commend the religion of the Europeans as something equally on a higher plane than any indigenous faith.

The first Jesuit residence, set up at Zhaoqing by Matteo Ricci, had a clock-face on the street with a public self-sounding bell, though this was closed down by a new Governor of the city in 1589. But modern horology was irresistible. A magnificent spring clock with chiming bells had been sent from Rome as a gift for the emperor, and arrived with the missionaries in the capital in 1600. After it had been installed in the imperial palace, the Jesuits were entrusted with its regulation, and the training of certain eunuchs in clock maintenance and repair. This was the beginning of nearly two centuries of service by the Jesuits, including lay-brothers trained as clock-makers, to the Chinese imperial court, where eventually there collected a great variety of clockwork instruments of all kinds. Moreover, wherever they established themselves in provincial cities their mechanical clocks were made known and appreciated. In sum, it is abundantly clear that one of the reasons why the early Jesuit missionaries were so much welcomed by the Chinese was for their interest in clocks and clock-making, hardly less indeed than for their skill as mathematicians and astronomers.

There can be no doubt that Matteo Ricci and his companions regarded efficient mechanical clocks as something absolutely new and unheard of in China. He says this in his memoirs on several occasions; thus the clock with three bells destined for the emperor, was a piece 'which struck all the Chinese dumb with astonishment, a work the like of which had never been seen, nor heard, nor even imagined, in Chinese history'. Ricci's opinion is unmistakable, and no other Jesuit thought differently as far as we know.

It is true that Ricci and Nicholas Trigault had something to say of Chinese clocks with driving-wheels which they found on their travels, though they laid little emphasis on them and their descriptions are obscure. It is also true that a number of contemporary Chinese scholars recognised that the clockwork of the Jesuit 'self-sounding' bells was not something fundamentally new in Chinese culture. The former were not anxious (in China) to exalt the achievements of the indigenous past. The latter were insufficiently learned to expound the past as it deserved. This complicated situation had its effects upon the thinking of Europeans later on, and in particular on European historians of science. If the Jesuits so firmly believed in the novelty of the mechanical clocks which they introduced to China, who were the later historians of science, penned within the ringed fence of the alphabetic languages, to gainsay them? This should indeed be a warning to all scholars of the provisional nature of their conclusions, and of the danger of preconceived ideas about the comparative contributions of the cultures of East and West.

What then was the accepted view, in closer detail? It held that the first successful achievement of slow, regular and continuous rotation in time with the daily revolution of the stars, by means of an escapement acting upon the driving-wheel of a train of gearing, occurred in Europe shortly after the beginning of the fourteenth century AD. When clocks are mentioned before that time, either sundials or water-clocks (clepsydras) are meant. The soul of the wheel-clock is the escapement, which hinders the rapid revolution of the wheels.

First let us gain a clear idea of the fourteenth-century mechanisms. The simplest form of the early European mechanical clock drew its power from the rotation of a drum brought about by the fall of a suspended weight. This was connected with trains of gearing in great variety, but the movement of the whole was slowed to the required extent by the escapement device known as the verge and foliot. Fig. 369 is a drawing of the essentials – the crown-wheel (II) with its teeth projecting at right-angles; the verge or rod standing across this wheel (KK) and bearing (at right angles to each other) two little plates or pallets so as to engage with the crown-wheel; and finally the foliot (LL) or 'crazy dancer', i.e. two weights carried one at each end of a bar set at the top of the verge. The twist of the crown-wheel, caused by the weight and rope acting on the drum (B) and the train of gears above it, pushed one of the pallets out of the way, giving a swing to the foliot. But this only led to the coming into action of the other pallet, which stopped movement of the crown-wheel. Its motion could only continue after the foliot and its weights had swung in the opposite direction. The other pallet was then moved out of the way by the crown-wheel. Thus the foliot oscillated back and forth, driven by the weight, at the same time imposing a step-by-step or ticking movement upon the gear train.

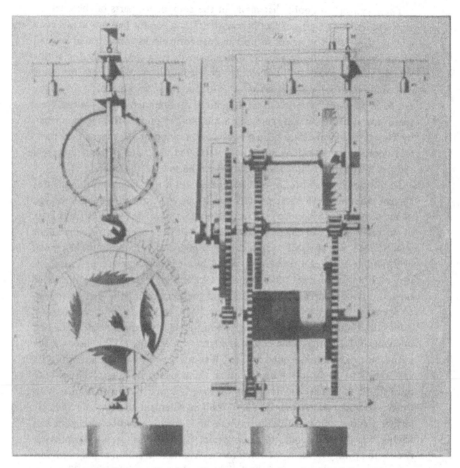

Fig. 369. Drawings to show the essentials of the verge-and-foliot escapement of early weight-driven mechanical clocks. Left, side view, right cross-section.

The more familiar examples of these early time-keepers are the large 'turret' clocks made for church towers and the like. Here a further complication is introduced, namely a striking gear train as well as the going train. The early forms of this are particularly interesting because the verge and pallets could act as a bell-ringing device if the swinging foliot weights were replaced by small hammers applied to a bell; the device would then, when released, run wild. It is very probable that the earliest European mechanical clocks had no pointer or dial-face but simply set off a striking arrangement when each hour, or other prearranged time, arrived.

There can be no doubt that many of the component parts of these clocks were Hellenistic in origin. Doubtless, the falling weight had originally been a falling float, such as we see in Roman anaphoric clocks, where a dial bearing astronomical markings was made to rotate slowly by a cord attached to a float sinking in a clepsydra. The idea of a dial would have come from the same source. Presently we shall see some evidence that this kind of system may not have been unknown in China, though never dominant there. In the West there was also the carriage carrying a mechanical puppet theatre; described by Heron of Alexandria, the necessary slow motion of the mechanical parts was obtained by the descent of heavy weights as grains of sand or cereal escaped through an hour-glass hole at the bottom of the container.

As for the verge-and-foliot escapement, it seems that Charles Frémont was correct when in 1915 he suggested that it derived from the radial bob type of flywheel (chapter 2, p. 57 above). Even so, one of the greatest mysteries of the early European clocks was the origin of the escapement principle. For a long time it was thought to appear in a strange design found in the notebook of Villard de Honnecourt about AD 1237, but it is now agreed that a mechanism such as this cannot have been an escapement, but simply a means of turning carved figures by hand. If so, no predecessor for the first European escapement remains – except the Chinese type shortly to be described.

For three hundred years the verge-and-foliot escapement itself remained unchanged, but towards the end of the sixteenth century AD the fact that the swings of a pendulum occurred in equal times began to attract attention. Its application to clocks in the next century seems primarily due to Galileo and, most of all, to Christiaan Huygens in 1657. At first it was combined with the verge and pallets, but in 1680 William Clement devised the familiar anchor escapement in which the crown-wheel was replaced by a scape-wheel having teeth in the plane of its rotation. This device persisted until the twentieth century in many modified forms. Another invention of major importance was the replacement of the failing weight by a spring-drive. This permitted the making of portable watches as well as stationary clocks.

Such was the general picture of the development of the mechanical clock as it stood on the basis of researches in European history alone. Certainly texts and remaining clock parts indicate that the earliest type was in use by about AD 1310, with the invention of the escapement occurring apparently at the beginning of the fourteenth century. As L. Bolton wrote in 1924: 'Weight-driven clocks come suddenly into notice at this period (fourteenth century) in a very advanced stage as regards design, though their workmanship was rough. Their previous evolution must have taken a long time, but there is no reliable record of its stages nor of the men responsible for it.'

In 1955, however, a way of solving this problem opened out to Dr Needham. For in the *Xin Yi Xiang Fa Yao* [*Hsin I Hsiang Fa Yao*] (New

Design for an Astronomical Clock), written in AD 1092 by a distinguished scientific scholar and civil servant of the Northern Song dynasty, Su Song [Su Sung], describes the erection in AD 1088 of elaborate machinery for effecting the measured slow motion of an armillary sphere and a celestial globe, together with a profusion of time-keeping jack-work (miniature human figures whose appearance indicated the time). The whole 'Combined Tower' which contained all this could in fact have been nothing more nor less than a great astronomical clock necessitating some form of escapement, as the full translation and study of the elaborate and detailed text showed. It also revealed the considerably earlier origins and development of time-keeping machinery recorded for posterity in Su Song's remarkable historical introduction. In this way six centuries of Chinese horological engineering, previously hidden, came to light.

SU ZIRONG [SU TZU-JUNG] AND HIS ASTRONOMICAL CLOCK

To readers of this abridgement Su Song (or Su Zirong) is by now a familiar figure. His work has been mentioned frequently in the section on astronomy (volume 2) in many connections. But he was not only an astronomer and a mathematician, he was also a naturalist, for in AD 1070 or thereabouts he produced, no doubt with a number of assistants, the best work of his age on pharmaceutical botany, zoology and mineralogy, the *Ben Cao Tu Jing* [*Pên Tshao Thu Ching*] (Illustrated Pharmacopoeia). Still today this treatise contains precious information on subjects such as iron and steel metallurgy in the eleventh century AD, or the use of drugs such as ephedrine, and we have often referred to it already, notably in the section on mineralogy (volume 2, pp. 311 ff). Su Song was primarily an eminent civil servant, but one of those (by no means few in medieval China) who mastered the scientific and technical knowledge of his time and found opportunities for employing it in the service of the state.

Su Song was born in AD 1020 in Fujian [Fukien], not far from Quanzhou [Chhüanchow], the city which Marco Polo was later to know as Zayton. He pursued his career in the bureaucracy with considerable success, associated neither with the Conservative party, though his friends were mostly members of it, nor with that of the Reformers; and he became a specialist in administration and finance. But as was usual in those days, he also received foreign assignments, and in 1077 was despatched as a diplomatic envoy to the Qidan [Chhi-tan] people of the Liao kingdom in the north. Ye Mengde [Yeh Mêng-Tê], in his *Shilin Yan Yu* [*Shih-Lin Yen Yü*] (Informal Conversations of (Ye) Shilin (Ye Mengde), tells how when he went on this particular mission, the opportunity came to Su Song to utilise his knowledge of astronomy and the calendar in which he had shown particular aptitude when under training:

... (Su Song) was sent as ambassador to offer congratulations (to the Liao emperor) on the occasion of his birthday, which happened to fall on the winter solstice. (At this time) our (Song) calendar was ahead of that of the Qidan [Chhi-tan] (Liao) kingdom by one day, and thus the assistant envoy considered that the congratulations should be offered on the earlier of the two days. But the secretary of protocol in the Qidan (Liao) Foreign Office declined to receive them on that day. As the (Liao) barbarians had no restrictions on astronomical and calendrical study, their experts in these matters were generally better (than those of the Song), and in fact their calendar was correct. Of course, Su Song was unable to accept it, but calmly and tactfully engaged in wide-ranging discussions on calendrical science, quoting many authorities which puzzled the (Liao) barbarian (astronomers) who all listened with surprise and appreciation. Finally he said that after all, the discrepancy was a small matter, for a difference of only a quarter of an hour would make a difference of one day if the solstice occurred around midnight, and that is considered much only because of convention. The (Liao) barbarians could not reject this argument, and Su Song was permitted to offer congratulations on the day desired (by his mission).

Upon his return he reported to the emperor Shenzong [Shen Tsung] who was very pleased and said that nothing could have been more embarrassing. When he asked which of the two calendars was right, Su Song told him the truth, with the result that the officials of the Bureau of Astronomy and Calendar were all punished and fined ...

At the beginning of the Yuanyou [Yuan-Ui] reign-period (AD 1086) the emperor ordered Su Song to reconstruct the armillary clock, and it exceeded by far all previous instruments in elaboration ... The original model was due to Han Gonglian [Han Kung-Lien], a first-class clerk in the Ministry of Personnel, who was a very ingenious man. By that time Su Song had become Vice-President (of the Chancellery Secretariat) and simply gave the ideas to him. He could always carry them out, so that the instrument was wonderfully elaborate and precise. When the (Jin [Chin]) barbarians captured the capital (Kaifeng) they destroyed the astronomical clock-tower (Fig. 370) and took away with them the armillary clock. Now it is said that the design is no longer known, even to the descendants of Su Song himself.

Fig. 370. Pictorial reconstruction of the astronomical clock tower built by Su Song and his collaborators at Kaifeng in AD 1090. The machinery with its water-wheel drive was fully enclosed in the tower; it rotated an observational armillary sphere on the top platform and a celestial globe in the upper storey. Time announcement was further fulfilled visually and audibly by numerous jacks mounted on the eight superimposed wheels of a timekeeping shaft appearing at windows in the pagoda-like structure at the front of the tower. The building was some 12 metres high. The staircase was actually within the tower. Original drawing by John Christiansen.

These concluding remarks will be elucidated hereafter; all that we need to note now is the high reputation of Su Song and his assistants in the subsequent generation.

Su's promotion to Vice-Minister took place some twelve years after his embassy and nearly twenty years after the appearance of his pharmacopoeia. It must have been partly the result of the success of the complete working wooden pilot model of the clock, which had been set up in the Imperial Palace at Kaifeng in the previous year (1088). In 1090 the sphere and globe were cast in bronze, and in 1094 the writing of Su Song's horological monograph was finished and presented. By this time he was seventy-five, the holder of many honourable titles, and one of the Deputy Tutors of the Heir Apparent. Dying in 1101, he did not see the tragedy of the fall of the capital two decades later and the flight of the Song empire to the southern provinces.

Our central point of interest now is his description of the power-drive of his armillary sphere, globe and jack-work, together with the escapement which controlled its movement. This is contained in the third chapter of his book, but first a word or two on the transmission of the text to us. Though available only in the north at the time of the preceding quotation, it was printed in the south in AD 1172. A copy of this edition was owned by the late Ming scholar Qian Zeng [Chhien Tsêng] (1629 to 1699) who reproduced it in a new edition with great care. It was printed again later on in the next century and more numerously in 1844. Another point of great interest is that it was by no means the only book on astronomical clockwork written during the Song dynasty. There was at least a book by Ruan Taifa [Juan Thai-Fa] entitled *Shui Yun Hun Tian Ji Yao* [*Shui Yun Hun Thien Chi Yao*] (Essentials of the (Technique of) Making Astronomical Apparatus Revolve by Water-Power), though nothing can be ascertained about this author, or his work, or date.

Su Song's general diagram of the works appears in Fig. 371, but its explanations may best be followed in the modern drawing of Fig. 372. The former sees the structure from the south or front, the latter from the southeast. The great driving-wheel, 3.35 metres in diameter (Fig. 373), carries thirty-six scoops on its circumference, into each of which in turn water pours at uniform rate from the constant level tank. The main driving-shaft of iron, with its cylindrical necks supported on iron crescent-shaped bearings, ends in a pinion which engages with a gear-wheel at the lower end of the main vertical transmission shaft. This drives two components. A suitably placed pinion connects it with the time-keeping gear-wheel which rotates the whole of the jack-work borne on the time-keeping shaft. This assembly (seen in the foreground of Fig. 371) consists of eight superimposed horizontal wheels, seven carrying round the jacks. Since each of these wheels is from 1.8 to 2.4 metres in diameter, the total weight involved must have been very consider-

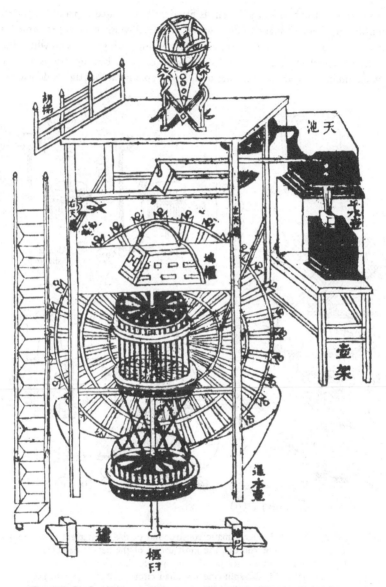

Fig. 371. Su Song's general diagram of the works. On the right, the upper reservoir tank with the constant-level tank beneath it. In the centre foreground, the 'earth horizon' box in which the celestial globe is mounted; below, the timekeeping shaft and wheels supported in the mortar-shaped end-bearing. Behind, the main driving-wheel with its spokes and scoops; above, the left and right upper locks with the upper balancing lever and upper link, curiously drawn, still higher.

able, so that the base of their shaft is fitted with a pointed cap and supported on an iron mortar-shaped end-bearing. The jack-work wheels performed a variety of functions, their figures either appearing with placards on which the time was marked, or ringing bells, striking gongs, or beating drums as they made their appearances in clothes of different colours at the pagoda doorways.

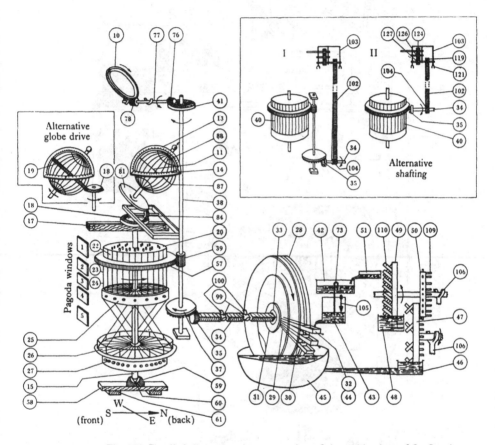

Fig. 372. Detailed diagrammatic construction of the mechanism of Su Song's clock tower. In our context, the most useful of the details depicted are: 10, the gear ring rotating once per day to operate the armillary sphere; 11, the celestial globe; 13, the split-ring meridian circle of the globe; 15, the timekeeping shaft; 20, wheel for striking the double-hours of the day by bells and drums; 22, 23, 24, 25, wheels with jacks for other divisions of time and for striking the night watches; 28, the great driving wheel; 42, upper reservoir tank; 43, constant-level tank; 48, intermediate water-raising tank; 47, lower water-raising wheel or noria; 49, upper water-raising wheel (for fuller details see Needham J., Wang & Price, *Heavenly Clockwork*, Cambridge, 1986).

Their rotation, however, was not the only duty of the time-keeping shaft, for at its upper end it engaged by means of oblique gearing and an intermediate idling pinion with a gear-wheel on the polar axis of the celestial globe (Fig. 372). The angle of these gears corresponded of course with the altitude of the celestial pole at Kaifeng. Now the text contains numbers of notes which record improvements in the clock, probably dating from the last years of the eleventh century, and in these an alternative globe drive is given, the uppermost gear-wheel rotating an equatorial gear-ring on the globe (Fig. 372). Possibly the original gearing proved difficult to maintain.

We must now return to the main vertical transmission shaft and the second component which it drives. Its upper end provides the power for the rotation of the armillary sphere. This is effected by right-angled gears and oblique gears connected by a short idling shaft. The oblique arrangement is made with a toothed ring called the 'daily-motion gear-ring' fitted round the intermediate nest or shell of the armillary sphere, not at the celestial equator of the sphere but at a 'declination' circle parallel to this and closer to the southern pole. Improvements were made as time went on. The vertical shaft was nearly 6 metres long and of wood. This must soon have showed itself to be mechanically unsound, and in the later variants (about AD 1100) it was first shortened and then abolished altogether. (See inset in Fig. 372.) The motive power was then conveyed to the armillary sphere by an endless chain-drive rotating three

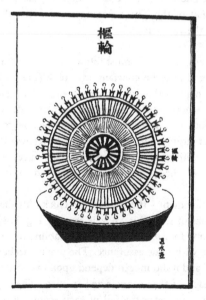

Fig. 373. Driving-wheel and sump from Su Song's *New Design for an Astronomical Clock* of AD 1092.

small pinions in a gear-box; see Fig. 284 (p. 72). This feature of the clock, may perhaps be considered the most remarkable of all for its time (the eleventh century AD), for although an endless belt of a kind had been incorporated in the magazine crossbow of Philon of Byzantium (third century BC) there is no evidence that this was ever built and certainly it did not transmit power continuously (See chapter 2, p. 71 above). A likelier source for Su Song's chain-drive may be found in the square-pallet chain-pump so widespread in the Chinese culture area, a device the origin of which we have traced back at least to the second century AD, and probably to the first. Of course this was for conveying material and not for transmitting power from one shaft to another – hence the originality of Su Song and his assistants, to whom perhaps indeed all true chain-drives are owing. Such is the interest of this feature that it may be worthwhile to give the description of it in the words of Su Song himself:

> The chain-drive (literally 'celestial ladder') is 5.9 metres long. The system is as follows: an iron chain with its links joined together to form an endless circuit hangs down from the upper chain-wheel which is concealed by the tortoise-and-cloud (column supporting the armillary sphere centrally), and passes round the lower chain-wheel which is mounted on the main driving-shaft. Whenever one link moves, it moves forward one tooth of the daily motion gear-ring and rotates the Component of the Three Arrangers of Time, thus following the motion of the heavens.

A brief description of the water-power parts must follow here. Water stored in the upper reservoir is delivered into the constant-level tank (Fig. 371) by a siphon and so passes to the scoops of the driving wheel, each of which has a capacity of 5.7 litres. As each scoop in turn descends the water is delivered into a sump. Apparently the clock was never so located as to be able to take advantage of a continuous water supply; instead of this the water was raised by hand-operated water-wheels with scoops or 'norias' in two stages to the upper reservoir. The bearings of these norias were supported on forked columns.

We can now examine the escapement – the soul of the clock. All that Su Song's draughtsmen could depict of it for his book is seen in Fig. 374, but fortunately the text is elaborate and for the most part clear, enabling the reconstruction of Fig. 375 to be made with some assurance. The whole mechanism was called the 'celestial balance' and it did indeed depend upon two steelyards or weighbridges upon which each of the scoops acted in turn. The first of these, the 'lower balancing lever', prevents the fall of each scoop until full, by means of a 'checking fork'. The basic principle is at once revealed, the

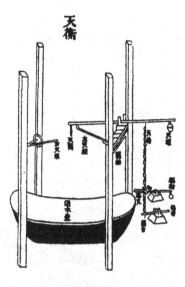

Fig. 374. The 'celestial balance' or escapement mechanism as given in Su Song's *New Design for an Astronomical Clock*. Legends in small characters, from left to right, upper row, read: right upper lock; upper link; left upper lock; axle or pivot; long chain; upper counterweight. Lower row: sump; checking fork of the lower balancing lever; coupling tongue; main (i.e. lower) counterweight.

determination of standard time units by the division of a constant flow of water into equal parts by a repeated process of accurate and automatic weighing by scoops carried on the driving-wheel. After each weighing operation the wheel is released so that it can make one step forwards under the power provided by the combined weight of several previously filled scoops (i.e. those between '3 o'clock' and '6 o'clock' on the wheel's rim, as seen in Fig. 371). Release then takes place. The whole sequence is best followed by examining Fig. 375 and its detailed accompanying caption.

There were 100 quarters (*ke*, [*kho*] 刻) in the Chinese day-and-night period of 12 equal double-hours, hence 8 1/3 quarters in each double-hour, not 8. As the *ke* was generally divided into 60 'minutes' (*fen*), this meant (1/3 × 60) or 20 extra *fen* in each double-hour. Thus the *ke* was equivalent to 14 min. 24 sec. of our time. To operate the jack-work at the correct moments in both the double-hours and the quarters, the machinery must have been designed to divide the day into both 24 and 100 equal parts. As the least common multiple of these numbers is 600, this explains why the main gear-wheels had just that number of teeth. (See Fig. 376.) The unit period was 1/600 part of a day, or 2 min. 24 sec. Close study reveals that the description in one of

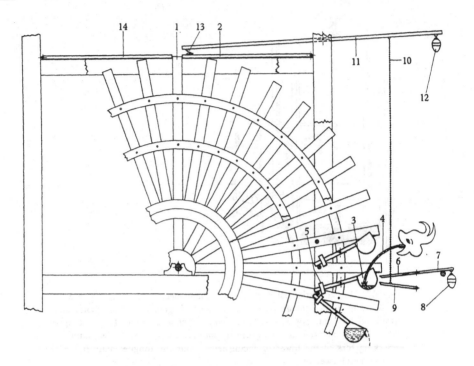

Fig. 375. Drawing of the escapement mechanism by John Combridge. 1, arrested spoke; 2, left upper lock (considered as the right in this analysis); 3, scoop, being filled by 4 which is a water jet from the constant level water tank; 5, small counterweight; 6, checking fork tripped by a projecting pin on the scoop, and forming the near end of 7, the lower balancing lever, with 8, its counterweight; 9, coupling tongue, connected by 10, the long chain with 11, the upper balancing lever, which has at its far end 12, the upper counterweight, and at its near end 13, a short length of chain connecting it with 2, the upper lock beneath it; 14 right upper lock, considered as left in this analysis.

At the beginning of each 24-second time interval the driving wheel is immobilised by the action of the right lock, 2, on the spoke 1. As water, 4, from the constant level tank enters the scoop, 3, the scoop holder counterweight, 5, is first overcome, and the excess weight of water then rests on the checking fork, 6, of the lower balancing lever 7. When the excess overcomes the counterweight, 8, the lever is suddenly tripped, and the scoop holder rotates about its pivot so as to fall sharply upon the coupling tongue, 9, and trip it in its turn. The long chain, 10, which passes freely between the prongs of the checking fork, 6, is thus abruptly pulled downwards, depressing the right hand end of the upper balancing lever, 11, with the aid of the counterweight 12, normally insufficient to effect this. Momentum is gathered from the loaded scoop for a brief instant while the levers swing, then the short chain of the upper link, 13, tightens and jerks the right upper lock, 2, out of the way of the spoke. The wheel

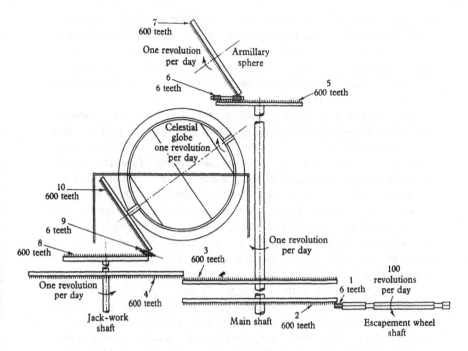

Fig. 376. Provisional drawing of the transmission machinery in Su Song's clock-tower. (See also Fig. 372.) Further study suggests that the 600-tooth gear wheels and the 6-tooth pinions were features of a preliminary experimental design. Su Song's actual clock-tower almost certainly had gear conversion mechanism from solar to sidereal rotation rates at the top of both shafts, as well as for the 'unequal' time intervals which changed with the lengths of daylight at different seasons.

Caption for Fig. 375. (*cont.*).

now makes one quick step clockwise under the driving force of the filled scoops in the lower right-hand quadrant, while the near or left-hand end of the upper balancing lever and the right-hand upper lock fall under their own weight again to stop the following spoke. Meanwhile the left-hand upper lock, 14, has been raised in ratchet fashion as the spoke has passed through, and now falls again behind the next spoke so as to prevent any recoil as the wheel stops. With the return of the linkwork to its original position, the levers 6, 7 and 9 regain their normal places ready for tripping in the following cycle. All the 'tick' processes are completed in an instant.

the texts that each movement of the driving-wheel corresponded to the passage of six gear-teeth of the timekeeping wheel, refers in fact not to the movement of one release step of the wheel, but to one complete revolution. This would give 36 ticks of 24 sec. each. Thus there were 6 ticks for each time-keeping gear-tooth passage, 36 in every quarter (*ke*), 300 in every double-hour, and thus 12 × 300 or 3600 every day.

The whole design is strangely reminiscent of the familiar anchor escapement of late seventeenth-century Europe, since there the driving-wheel is also a scape-wheel and the 'pallets' are inserted alternately at two points on its circumference separated by 90° or less, rather than the 180° of the crown-wheel. Although the Chinese solution of the problem by chain and linkwork has a certain medieval cumbrousness, the operation is elegant and the performance accurate to an unexpected degree. It certainly far exceeded the inventive capacity of contemporary Europe, that other culture-area where a purely mechanical escapement would later appear. In the Chinese water-wheel linkwork device the action of stopping and releasing the driving- or scape-wheel is brought about not by the oscillation under gravity of a pendulum as in the anchor escapement, but by the force of gravity exerted periodically as a continuous steady flow of liquid fills containers of limited size. This type of escapement had remained quite unknown to historians of technology until the elucidation of Su Song's text. Its peculiar interest lies in the fact that it constitutes a 'missing link' between the time-measuring properties of liquid flow and those of mechanical oscillation. It thus unites, under the significant sign of the millwright's art, the clepsydra or water-clock and the mechanical clock in one continuous line of evolution.

At one point in the text it is said that one of the jack-wheels strikes a gong to indicate the night-watches as it turns. All such auditory performances of the jacks must have involved simple contrivances of springs, probably of bamboo. Hence the interest of the statement made in the following century by Xue Jixuan [Hsüeh Chi-Hsüan]: 'Nowadays time-keeping devices are of four different kinds. There are the clepsydras, the (burning) incense stick, the sundial, and the revolving and snapping springs.' This last expression seems to be rather a rare one, absent both from the copious engineering vocabulary of Su Song himself and from the mass of other texts which concern the development of clockwork in medieval China. The text is a notable one, for it was adduced in the time of the Jesuits by the few Chinese scholars who knew enough on such subjects in those days to point out that their Renaissance clocks were not the first which had been known in China.

This completes the account of the hydro-mechanical clockwork of Su Song's great astronomical tower, set up in the form of a working wooden pilot model in the imperial palace at Kaifeng in AD 1088. This was the time of our Doomsday Book and the youth of the philosopher, poet and theologian Peter

Abelard. Two years later the metal parts, i.e. the armillary sphere and the celestial globe, were duly cast in bronze. The writing of the explanatory monograph must have been well under way in 1092, and it was finally presented to the throne in 1094. Prefixed to it is a remarkable memorial in which Su Song not only describes the principles of the clock itself, but gives a historical disquisition on all instruments of a similar kind which had existed in previous centuries. This it was which illuminated many other texts not previously comprehensible, permitting the establishment of a history of Chinese clockwork, the outline of which follows. Moreover, it describes the organisation of what is one of the greatest technical achievements of the medieval time in any civilisation. Reading 'between the lines' also brings out several other significant points. Han Gonglian [Han Kung-Lien], a man of brilliant mathematical and mechanical talent, who helped oversee the construction, was found a post by Su Song so that he could make full use of his abilities. Then contrary to common conceptions of medieval working, the new armillary clock was not put together haphazardly by trial and error, but planned in a special memorandum with all the geometrical knowledge which Han could muster. This certainly makes it easier to understand how the gearing, chain-drives, and other devices were made to carry out successfully their duty of rotating steadily an armillary sphere weighing some 10–20 tonnes as well as a bronze celestial globe some nearly 1.4 metres in diameter. It is also noteworthy that a small wooden model was first made, then a full-scale one was tested against four types of clepsydra as well as star transits, and only after four years were the parts destined for bronze duly cast.

In the last paragraph of his memorial Su Song wrote:

> Thus, as we have seen, the (demonstrational) armillary sphere, the bronze observational armillary sphere, and the celestial globe, are three things different from one another . . . (therefore) if we use only one name, all the marvellous uses of (the three) instruments cannot be included in its meaning. Yet since our newly built machine embodies two instruments but has three uses, it ought to have some (more general) name such as 'Cosmic Engine'. We are humbly awaiting your Imperial Majesty's opinion and bestowal of a suitable name upon it.

Clearly the clock's uses were demonstrations of astronomical motions and astronomical observations, both with the armillary sphere, together with an indication on the globe of the positions of all constellations whatever the weather, and their relations to the sun, moon and planets attached to the globe for verifying the calendar. Besides these functions there was also the indication of time, both visually and by sound, using elaborate jack-work. Thus Su

Song's request for a new name was of great historical significance. The mechanised astronomical instrument was trembling on the verge of becoming a purely time-keeping machine. Inaudible echo must have answered 'A clock!' But history does not record that the young emperor had any good ideas on nomenclature, and the time-measuring function continued to go unnamed until five hundred years later the Jesuits came with their 'self-sounding bells' to ring in the age of unified world science with its unlimited expansion of appropriate technical terms.

CLOCKWORK IN AND BEFORE THE NORTHERN SONG

The derivation of mechanical clocks from clepsydras has now been made clear. But the story of the evolution remains to be told. In his memorial Su Sung wrote:

> According to your servant's opinion there have been many systems and designs for astronomical instruments during past dynasties all differing from one another in minor respects. But the principle of the use of water-power for the driving mechanism has always been the same. The heavens move without ceasing but so also does water flow (and fall). Thus if the water is made to pour with perfect evenness, then the comparison of the rotary movements (of the heavens and the machine) will show no discrepancy or contradiction; for the unresting follows the unceasing.

This was a nice appreciation of what Europeans were later to think of as the universal writ of the 'law' of gravitation. But Su Song goes on to give brief descriptions of the predecessors of his own clock, beginning with the second century AD device of Zhang Heng [Chang Heng]. In this we shall follow him, making use of the original texts on which his summary was based. The most convenient plan will be to begin by working backwards, starting from the eleventh century and going back as far as the first century AD. We shall then be able to make another start from his time in the opposite direction, and following the fate of his own machine, describe the chief events in Chinese clock-making which took place between the twelfth century AD and the arrival of the Jesuits at the end of the sixteenth.

We must first take up an important mechanical point. It will have been noticed that the description of Su Song's clock contains nothing resembling a dial. Though the stationary dial-face with a moving pointer is a development associated with the first European mechanical clocks of the fourteenth century AD, as we have seen (p. 224), the rotating dial-face had been used in Roman times. Such an anaphoric clock, described by Vitruvius about BC 30, consisted of a bronze disc on which there was a planisphere depicting the stars of the northern hemisphere, and as many as are found between the equator and the tropic of Capricorn which formed the rim of the disc. The circle representing the ecliptic (the zodiac) was provided with 365 small holes, into which was plugged from day to day a little stud representing the sun. We note again that the disc was made to rotate by the simple mechanism of a float in a clepsydra attached to a cord. This terminated in a counterweight and wound round a drum on the horizontal axis bearing the disc with the planisphere. In Alexandria the disc of such clocks was separated from the spectator by wires which depicted the meridian, the equator and the tropics and the months measured by the zodiacal constellations, as well as the time, while the horizon was also indicated. This has been shown to be the forerunner of the astrolabe yet the astrolabe was never known in China. The question does arise, however, whether any similar kind of mechanised planisphere was employed there. Apart from the Vitruvian description, actual bronze engraved discs from the Roman Empire have survived in fragmentary form. No such objects have yet come to light in China, or at least none such has so far been recognised.

However, there are certain literary evidences which suggest that simple anaphoric clocks were not unknown there. In the *Dong Tian Qing Lu* [*Tung Thien Chhing Lu*] (Clarifications of Strange Things) of about AD 1240, Zhao Xigu [Chao Hsi-Ku] says that a certain Fan Zhongyan possessed 'an ancient mirror', on the back of which the twelve double-hours were marked by protuberances, and that whenever such an hour arrived, one of the protuberances shone like the full moon. The whole device rotated ceaselessly. Another earlier scholar said that there was a 'Twelve (Double-) Hour Bell' which sounded automatically as the hours passed. These devices would have existed about AD 1020. Yet about AD 950 Tao Gu [Thao Ku] tells in his *Qing Yi Lu* [*Chhing I Lu*] of a mirror which had been in use at the beginning of the ninth century AD. This had a 'yellow plate' with a circumference of almost one metre, around which were designs of animals and other things. At each double-hour one of these appeared. It was called the 'Twelve (Double-) Hour Plate' and was still in existence at least until AD 923. However, Tao Gu does not distinctly state whether or not the turning was automatic, but this seems probable. An even earlier account seems to show the beginnings of timekeeping jack-work, the figures making their rounds on a horizontally mounted wheel like those of Su Song's. This is in *Chao Ye Qian Zai* [*Chhao Yeh Chhien Tsai*] (Stories of Court Life and Rustic Life) written Zhang Zu [Chang Tsu] early in the eighty century AD. The device – a 'Wheel (for Reporting) the Twelve (Double-) Hours' – dated from AD 692, and was made therefore several decades before the first escapement clock, that of Yixing [I-Hsing] (see below, p. 241). Since the effects it produced would not have required much

power to drive, it is reasonable to assume a sinking-float mechanism. Not long
before (about AD 500) a celestial globe had been rotated in India by means
of the anaphoric clock principle, though evidence for this comes from a
fifteenth century source.

The sinking-float principle never seems to have been prominent in the
Chinese culture-area. This was probably because from the outset, as we shall
see, it was desired to rotate not only discs with planispheres but spherical
astronomical instruments, which even if made of wood were quite heavy. The
power requirements necessitated the enlistment of the water-wheel – and
indeed, this was done probably within a century of the first appearance of
the water-mill itself. Our immediate hunt is for the first appearance of the
escapement.

The most important clock in the Song dynasty prior to that of Su Song
himself was built by Zhang Sixun [Chang Ssu-Hsün], towards the end of the
tenth century AD. It included a sphere and globe, powered by a scoop-
bearing driving-wheel and gearing, together with jack-figures to report and
sound the hours. Eleven technical terms occur in the description with exactly
the same meanings as in Su Song's text. Zhang's clock was a particularly fine
and interesting work as it later used mercury in the closed circuit instead of
water, thus assuring timekeeping in frosty winters. But it must have been
somewhat ahead of its age, for Su Song tells us that after Zhang's death it
soon went out of order and there was no one to keep it going. His description
also makes it clear that it was very like his own clock. From this it also appears
that Zhang's clock possessed a gear-box, which rather implies the use of a
chain-drive like Su Song's; if so Zhang Sixun was an anticipator of Leonardo
da Vinci by five hundred years.

From this we can pass directly to the 'Tang clockmakers'. Who were
these men who made, in the eighth century AD, the most venerable of all
escapement clocks? One was a Tantric Buddhist monk, perhaps the most
learned and skilled astronomer and mathematician of his time, Yixing
[I Hsing], the other a scholar, Liang Lingzan [Liang Ling-Tsan], who occupied a minor administrative post. The technical terms employed in the relevant passages again reveal the essential similarity of the machine to the clock
of Su Song.

The passages are to be found in the official histories of the Tang dynasty
and in the *Ji Xian Zhu Ji* [*Chi Hsien Chu Chi*] (Records of the College of
All Sages) written about AD 750 by Wei Shu. From all these we find that
in AD 723 Yixing and Liang Lingzan 'and other capable technical men' were
commissioned to cast and make new bronze astronomical instruments.

One of these was a celestial sphere showing the lunar mansions in order,
and the celestial equator and the degrees 'of the heavenly circumference'. The
text of the Tang history goes on:

Water, flowing (into scoops), turned a wheel automatically, rotating it (the sphere) one complete revolution in one day. Besides this there were two rings fitted round the celestial (sphere) outside, having sun and moon threaded on them, and these were made to move in circling orbit . . . And they made a wooden casing the surface of which represented the horizon, since the instrument was half sunk in it. This permitted the exact determination of the times of dawns and dusks, full and new moons, tarrying and hurrying. Moreover there were two wooden jacks standing on the horizon surface, having one a bell and the other a drum standing in front of it, the bell being struck automatically to indicate the hours, and the drum being beaten automatically to indicate the quarters.

All these motions were brought about (by machinery) within the casing, each depending on wheels and shafts, hooks, pins and interlocking rods, coupling devices and locks checking mutually (i.e. the escapement).

Such are the details of the instrument which, so far as we can see, was the first of all escapement clocks. The reference to checking linkwork is plain, the technical terms used being closely similar to those in the descriptions of the clock of Su Song. Although the automatic movement of the sun and moon models is not stated with absolute clarity, it is almost certainly implied; the machine had therefore some at least of the features of an orrery or planetarium.

For its background we have to look at the preceding reign. The emperor was Taizong [Thai Tsung], who ruled with much brilliance from AD 626 onwards for a quarter of a century. Interested in history and technology as well as in the military arts, he knew how to encourage astronomers, and welcomed Nestorian clergy as well as Daoist priests and Buddhist monks. He entertained cordial diplomatic relations as far west as Byzantium, and from missions which visited him he may have heard of striking water-clocks at places like Gaza and Antioch. Of course this can have been no more than 'stimulus diffusion', for there is no reason for thinking that the Byzantine works employed anything more than the sinking-float principle. However, the stimulus would have come at just the right time to encourage Chinese engineers to try to outdo the mechanical toys which formed the striking jack-work of the water-clocks of the Eastern Roman Empire. And indeed the description of Yixing's clock does seem to be the first mention of horological escapement-operated jacks in Chinese history. Here he was much better placed than his Greek colleagues, if it is correct to suppose that the water-wheel, providing so much more power than the float, had already long before been characteristic of Chinese astro-mechanical technique.

The emperor for whom Yixing worked was Xuanzong [Hsüan Tsung], most unfortunate among rulers of the Tang. Ascending the throne in AD 712, he prospered for some thirty years, shining as a patron of music, painting and literature. All the greatest Tang poets knew his court. In later life, however, growing social and economic strains exposed the country to the military rebellion of An Lushan, from which the dynasty never recovered. Among its incidental results was the death of Xuanzong's famous and beautiful concubine Yang Guifei [Yang Kuei Fei]. Though Yixing's clock cannot have been built for her, since she did not join the imperial entourage till 738, her presence evokes considerations of a rather singular kind.

It will be recalled that the Chinese emperor was a cosmic figure, the analogue here of the pole star on high. All hierarchies, all officialdom, all works and days, revolved around his solitary eminence. It was therefore entirely natural that from time immemorial the large number of women attending upon him should have been regulated according to the principle of sacred cosmic outlook which pervaded Chinese court life. Though their titles differed considerably during the two thousand years which followed the first unification of the empire, the general order comprised one Empress, three Consorts, nine Spouses, twenty-seven Beauties (concubines), and eighty-one Attendant Nymphs. Women of the highest rank approached the emperor at times nearest to full moon, when the Yin influence would be at its height, and matching the powerful Yang force of the Son of Heaven, would give the highest virtues to the children so conceived. The primary purpose of the lower ranks of women was rather to feed the emperor's Yang with their Yin. Secretaries kept a careful record of all this, and it is their activities which show us the relevance of these curious matters to the invention of clockwork.

What was at stake was the imperial succession. Chinese ruling houses did not necessarily follow the principle that the eldest son of the empress was necessarily the heir apparent. Towards the end of a long reign the emperor would have quite a number of princes from which to choose, and in view of the importance of state astrology in China from very ancient times it may be taken as certain that one of the factors in the choice was the nature of the star-groups reaching greatest height in the sky at the time of the candidate's conception. Hence the importance of the records kept by the 'Duennas Secretarial', and the value of an instrument which not only told them the time but from which the eunuchs could read off the star positions at any desired moment. Interest in recording machinery is therefore not at all surprising. But these facts, like the interest in time-keeping which monks in the West had in order to regulate their set times of daily prayer, fall short of giving us the reason why the invention of the escapement, as the reply to a demand for more accurate time-keeping, should have come about particularly at this juncture.

Before quitting the subject of these earliest escapement clocks it is necessary to say a little more about certain inventions which paved the way for them. For instance, we have found indications of horizontal jack-wheels in the century preceding Yixing. It will also have become clear that in the absence of oscillators (such as the verge-and-foliot or pendulum) an essential part of the water-wheel linkwork escapement was formed by the two weighbridges or steelyards which the scoops had to trip. To understand where this component could have come from, we must glance at the history of water-clocks in ancient and medieval China.

The most archaic sort of water-clock, the outflow type, was doubtless a gift to East Asia from the cultural centres of the Fertile Crescent – Lebanon, Syria and down to the Persian Gulf – in the first or even second millennium BC. But from about 200 BC onwards it was replaced almost everywhere in China by the inflow type with an indicator-rod carried on a float. Already in the Han it was well understood that a falling head of pressure in the reservoir tank greatly slowed the time-keeping as the inflow vessel filled. Throughout the centuries two methods of correcting for this were adopted, first interposing one or more compensating tanks between reservoir and inflow vessel, and then a little later inserting an overflow or constant-level tank in the series. But there were others which involved weighing the water on some kind of balance.

Balance water-clocks included at least two types, one in which the steelyard was applied to the inflow vessel itself, and another in which it weighed the amount of water in the lowest compensation tank. The latter was used for public and palace clocks throughout the Tang and Song periods. It permitted seasonal adjustment of the pressure-head in the compensating tank by having standard positions for the counterweight graduated on the beam, and hence it could control the rate of flow for different lengths of day and night. No overflow tank was required, and attendants were warned when the water-clock needed refilling. Designed in the Sui dynasty by Geng Xun [Kêng Hsün], one of the most outstanding technicians of the sixth century, whom we have already met on account of his work on water-clocks (volume 2, p. 155 of this abridgement) and by Yuwen Kai [Yüwên Khai], this method was established just about AD 610, i.e. one hundred and ten years before the work of Yixing and Liang Lingzan.

The relevance of these facts is obvious. From the systematic weighing of a water-receiving vessel it was not such a far cry to the weighing of one which both received it and delivered it. That in turn would have pointed the way to mounting such double-function containers round the rim of a single wheel; in fact to the controlled retention of successive water-receiving and water-delivering scoops so arranged as to pass a weighbridge.

THE PREHISTORY OF CHINESE CLOCKWORK

We have concluded, then, that what may be called the father-and-mother of all escapements originated in the first decades of the eighth century AD. But the history of clockwork which we are sketching cannot stop at this point on its backward course. For between AD 725 and the beginning of the Christian era there were many other examples of astronomical globes or spheres being slowly rotated by water-power. If we define the escapement as the essence of true clockwork, these earlier devices were not exactly clocks, but they may well be considered the predecessors of the clock. Moreover, they were purely astronomical in character and lacked gongs and drums for telling the time.

Since the meagre power of the float could not have rotated such comparatively heavy inclined globes, even if made of wood, it is possible that the mechanism consisted of a vertical water-wheel with cups, doubtless more simple in construction than that of Su Song. The water-wheel could be attached to a shaft in the polar axis (the axis of rotation of the inclined globe) with one trip-lug, quite similar to the principle of the water-driven trip-hammer assemblies so common in the Han. Water-clock drips into cups would accumulate periodically the torque necessary to turn the lug against the resistance of a leaf-tooth wheel, either itself forming the ring representing the equator, or attached to the shaft of the polar axis. Needless to say, the timekeeping properties of such an arrangement would be extremely poor, so poor indeed that it is hard to understand how they could have justified the explicit claims that have come down to us in the ancient texts. Perhaps the linkwork escapement is older than we have dared suppose.

Examples may be taken from almost every century between the eighth and second AD. One of the most outstanding technicians of the sixth century was Geng Xun [Kêng Hsün] and seventh-century biography of him, it is said:

> ... Xun then conceived the idea of making an armillary sphere which should be turned not by human hands but by the power of (falling) water. When it had been made he set it up in a closed room and asked (Gao) Zhibao [(Kao) Chih-Pao] (his old friend the Astronomer Royal) to stand outside and observe the time (as shown by the) heavens (i.e. the star transits). (His instrument) agreed (with the heavens) like the two halves of a tally ...

This account, referring to work which was going on in the neighbourhood of AD 590 is closely similar to what we are told in all the other cases.

Some seventy years earlier, about AD 520, the great Daoist physician, alchemist and pharmaceutical naturalist, Tao Hongjing [Thao Hung-Ching] (AD 452 to 536), had done something of a similar kind. He also wrote a book about it, which has since been lost. A century earlier still, under the Liu Song dynasty, similar apparatus was set up by Qian Lozhi [Chhien Lo-Chih], an astronomer who was probably the originator of solid celestial globes in China. His work has the special interest that it followed the recovery of the old astronomical instruments of Zhang Heng [Chang Hêng] (AD 78 to 139). Even more significantly, Qian's instruments lasted a long time; they were almost certainly known by Geng Xun and Yuwen Kai two centuries later, and by Yixing and Liang Lingzan a hundred years after that, in AD 723.

Another intermediate figure of whom we know, however, was Ge Heng [Ko Hêng]. In the San Guo [San Kuo] State of Wu (AD 222 to 280):

> . . . there was also Ge Heng who was a perfect master of
> astronomical learning and capable of making ingenious apparatus.
> He altered the astronomical instrument in such a way as to show
> the earth fixed at the centre of the heavens, and these were
> made to move round by a mechanism while the earth remained
> stationary. (This demonstrated) the correspondence of (the shadows
> on the) graduated sundial with the motions (of the heavens above).
> This it was which Qian Lozhi (also) imitated.

We are now within a century of our goal, the work of Zhang Heng [Chang Hêng] in the Later Han period. For he was the first of this line of men who accomplished the continuous slow rotation of astronomical instruments (globes or spheres for demonstration) with the best approximation they could make to constancy of speed. To readers of this abridgement Zhang Heng is a familiar figure; in volume 2 there was hardly a section where his name did not appear (mathematics, astronomy, map-making, etc.). Particularly relevant is the seismograph which he set up at the capital in AD 132, the first instrument of the kind in any civilisation. The ingenuity of this device, with its inverted pendulum, was so striking that there is nothing inherently improbable in his application of water-power to a drive for an astronomical instrument.

Our two most explicit sources both date from the Tang, though compiled of course on the basis of ancient documents then existing. One comes from the *Sui Shu* (History of the Sui Dynasty) of about AD 656, and the second from the *Jin Shu* [*Chin Shu*] (History of the Jin Dynasty), two works which stand out prominently among official histories for the length and excellence of their chapters on astronomy and science of the calendar. The *Sui Shu* says:

> . . . The Astronomer Royal Zhang Heng again (cast) a bronze
> instrument, on the scale of 1 cm to the degree, its circumference
> being 4.45 metres. It was placed in a closed chamber and rotated
> by the water of a clepsydra (literally, dripping water). One observer

watched behind closed doors and called out to another observer
who was looking at the heavens on the observatory platform, saying
when such and such a star should be rising, or making its transit,
or setting, and everything corresponded like the two halves of a
tally.

The account in the *Jin Shu* is similar, though it goes into more astronomical
detail about the rotating instrument itself. However, its last sentence is indeed
significant:

> The transits, risings and settings of the heavenly bodies (shown on
> the instrument in the chamber) corresponded with (literally,
> resonated with) those in the (actual) heavens, following the trip-lug
> and the turning of the auspicious wheel.

The observational procedure is clear, but as to the machinery employed, only
this last sentence gives a clue.

It is unsatisfactory that our chief sources date from two to five centuries
after Zhang Heng himself. However, contemporary evidence does exist, and
four pieces may be mentioned. First, two later writers quote directly from
his own books, which in their time still existed. Thus Su Song wrote in his
memorial to the emperor (AD 1092): 'Zhang Heng . . . says that (one instrument) should be set up in a closed room and rotated by water-power . . .' And
much earlier, about AD 750, Wei Shu wrote of Yixing's clock that when
people discussed it they said that what Zhang Heng described in his *Ling Xian*
[*Ling Hsien*] (Spiritual Constitution of the Universe) could have been no
better.

Besides this, two fragments of Zhang Heng's concerning water-clock technique were preserved by Tang writers, but their content is not relevant here.
However, the important thing about them is that they provide evidence that
the claim to have succeeded in constructing a machine of this kind seems to
be Zhang Heng's own and not something fathered on him by later generations.

The fourth piece of evidence is the most interesting, for though telling us
nothing of the mechanism, it reveals the whole principle and procedure for
which it was devised. It is a passage in one of the Later Han apocryphal books,
the *Shang Shu Wei Kao Ling Yao Yi* [*Shang Shu Wei Khao Ling Yao I*]
(Apocryphal Treatise on the Historical Classic; the Investigation of the
Mysterious Brightnesses). Its most probable date is about contemporary with
Zhang Heng. It runs as follows:

> If the (demonstrational armillary) sphere indicates a meridian transit
> when the star (in question) has not yet made it, (the sun's apparent
> position being correctly indicated), this is called 'hurrying'. When

'hurrying' occurs, the sun oversteps his degrees, and the moon does not attain the *xiu* [*hsiu*] (mansion) in which she should be. If a star makes its meridian transit when the (demonstrational armillary) sphere has not yet reached that point, (the sun's apparent position being correctly indicated), this is called 'dawdling'. When 'dawdling' occurs, the sun does not reach the degree which it ought to have reached, and the moon goes beyond its proper place into the next *xiu*. But if the stars make their meridian transits at the same moment as the sphere, this is called 'harmony'. Then the wind and rain will come at their proper time, plants and herbs luxuriate, the five cereals give good harvest and all things flourish.

With this it is interesting to compare the comments of Su Song, who quoted it towards the end of his memorial.

From this we may conclude that those who make observations with instruments are not only organising a correct calendar so that good government can be carried on (i.e. the administration of agricultural society), but also (in a sense) predicting the good and bad fortune (of the country) and studying the (reasons for) the resulting gains and losses.

Thus he rationalises the common prognosticatory significance attributed by lay folk to the activities of astronomers in medieval China. Not astrological presage but sound calendrical science will make a country prosperous, he says.

Here then is the key to the apparently mysterious arrangement of having one sphere in a closed room and another on the observatory platform. The calendar-making astronomers of Zhang Heng's time were very much concerned with all divergences or discrepancies between the indicated positions of the stars on the one hand and those of the sun and moon on the other.

The elucidation of this procedure gives us another opportunity of studying the social context of the inventions of the 'pre-clock' (if we may so term Zhang Heng's device) and the escapement wheels. The promulgation of the official calendar was one of the most important acts of the Chinese emperor, and about one hundred of these were issued from the first unification of the empire in the third century BC until the end of the Qing dynasty in the nineteenth century AD. Each bore a specific name. The question may now be asked whether there was any relation between the horological inventions and the frequency of introduction of new calendars. Were the inventions connected with what one might call periods of 'calendrical uneasiness'?

A preliminary answer is not at all difficult. Problems were beginning to be posed in the Han, and had mostly been solved by the Qing. Zhang Heng's invention occurs in the preparatory Han period, at a time of few calendars, but one must remember that he was setting in motion a technique which would have to be followed for many years to bring successful results. On the other hand, the indubitable use of the escapement by Yixing comes just towards the latter part of the burst of 'calendrical uneasiness' and calendar-making activity of the Liu Chao and Tang periods. The great clocks of later times all come within the similar period of the Song.

If the evidence is plotted from AD 200 to AD 1300 in 25-year periods (Fig. 377), the phases acquire considerable clarity. Up to AD 50 there is relative quiet, but from then onwards (apart from a spate of inaugurations) there is at least one calendar every quarter-century. At the beginning of the eighth century the merits of other calendars – from India, Persia and Uzbekistan – were hotly debated at the Chinese capital. This was the very time that Yixing and Liang Lingzan, as if in answer to desperate demands for some more truly timekeeping machine, made their invention of the water-wheel linkwork escapement. Towards the end of the Tang things were again quieter, but with the Song discrepancies must have become obvious once more and we find the clock of Zhang Sixun (AD 975) coinciding with a quarter-century in which no less than three calendars were produced. Between 925 and 1225 no quarter-century passed without at least one new calendar, so the clock of Su Song (1088) finds its place naturally in this period. Then in the last part of the thirteenth century there is a burst of activity which is readily explained by Arabic influence, like that of the Indian and Persian. Eventually comes the lethargy of the Ming, and the beginnings of unified world astronomy with the Jesuits in the early Qing.

Thus side by side with considerations arising from the private life of the imperial family we can place the need for computing calendars. It looks as if the invention of Zhang Heng about AD 120 derived from the growing doubts which had led Hipparchus in Greece to the discovery of the precession of the equinoxes in 134 BC and were to lead Yu Xi [Yü Hsi] to the same doctrine about AD 300. In this abridgement it has often been emphasised that Chinese astronomy was founded and built upon a system of coordinates based on the celestial pole and the celestial equator (see volume 2 of this abridgement, p. 117), while that of the Greeks was primarily ecliptic and planetary. Each had its peculiar advantages and its corresponding triumphs. If the discovery of Hipparchus preceded that of Yu Xi by four and a half centuries, it was because he was using a system which made evident positions of stars with respect to the equinoxes. But if astronomical instruments were rotated mechanically by Zhang Heng fifteen centuries before the conception of the clock-drive arose in Renaissance Europe, and if in this Yixing with his well documented and successful mechanised time-keeping had high priority, it was because Chinese astronomers thought always in terms of equatorial coordinates. In China therefore it was an entirely natural thought to arrange the

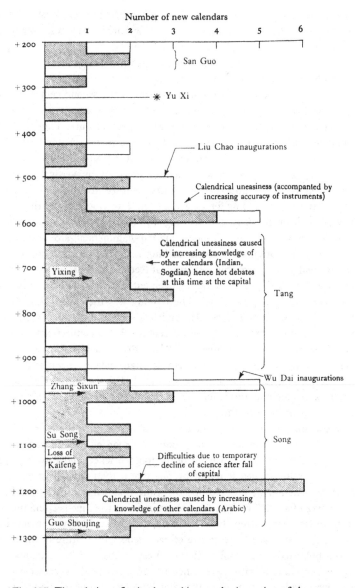

Fig. 377. The relation of calendar-making to the invention of the
escapement; plot of the number of new calendars introduced in each
25-year period between AD 200 and 1300. Lines bounding shaded areas
indicate new calendars other than those marking the beginning of new
dynasties. Thinner lines, mostly bounding the light areas, give the number
of new calendars for each 25-year period (excluding mere changes of name).

rotation of a celestial globe or armillary sphere used for demonstration if the plan promised to be useful. What was not perhaps so easy was how to do it.

This discussion may end, then, with a few words to set against their proper background the means which Dr Needham thinks Zhang Heng took to solve his problem. To harness the waste dripping of clepsydra water was one obvious recourse, the art of the millwrights. In the second century AD their workmanship was doubtless primitive, but during the previous century the water-powered trip-hammer had come into widespread use. Though the horizontal mill-wheel was the most prevalent type in China, the vertical kind always persisted, and for the trip-hammer was the more suitable of the two. Moreover, metallurgical blowing-engines powered by vertical wheels had become common in China during the first century AD, so that Zhang's chief originality lay in arranging for a constant drip into scoops rather than a strong flow and fall on paddles. The trip-lug on his shaft merely corresponded to those which worked the grain-pounding trip-hammers all around him. However, there was originality also in making it push each time the tooth of a ring or gear-wheel, probably bearing leaf-teeth, and controlled by a ratchet. There seems nothing at all in the arrangement which would have been beyond the powers of Han technicians. The trip-lug was their own, though to use it as a pinion of one was rather Alexandrian, such as with the peg on the axle of Heron's hodometer in the previous century. What connections there could have been between the engineers of Alexandria and Han China remains of course a completely unsolved problem. It might have been a case of carrying coals to Newcastle, for as we have seen (p. 184 ff above) the Chinese hodometers were contemporary, and their acceptable reconstructions involve small pinions of one, two or three teeth.

Presumably Zhang's simple machine was at rest during each period when water accumulated in one of the scoops. As soon as enough had collected, its weight overcame the resistance of the toothed wheel and armillary sphere, and the trip-lug turned it round by one tooth, then coming to rest against the next. Though we are told optimistically that 'everything agreed like the two halves of a tally' we are bound to assume that the time-keeping properties of the device were extremely poor. So much of it would have depended upon play and resistance, the exact size of each scoop, the nature of the bearings of the polar axis, and similar factors. Very probably it was maintained in regular motion only with some difficulty, and perhaps the successive astronomers who made instruments of the same kind were all searching for the right conditions for doing what no one until Yixing was really able to do. Or perhaps they made the machine work better than one is inclined to believe. In any case it is hard to exclude Zhang Heng's apparatus absolutely from the definition of von Bertele with which this chapter began. The problem of securing slow and regular motion by intersecting the progress of a powered machine into equal intervals was not solved by any Alexandrian design as far as is known, and if Zhang Heng did not solve it himself, he opened the gate to the path which led in the end to its solution.

FROM SU ZIRONG [SU TZU-JUNG] TO LI MADOU [LI MA-TOU]; CLOCKS AND THEIR MAKERS

We now return to our focal point, the great clock built by Su Song in AD 1088, in order to start again from there and trace briefly the subsequent developments down to the time of the coming of Matteo Ricci (Li Madou) and his Jesuit horologists. This is a space of almost 500 years, and contrary to all former belief, they were filled with clock-making in China, the success of the mechanicians varying with time and place.

The most obvious question to begin with is what fate befell Su Song's own clock-tower and the machinery inside it. It so happens that we are rather well informed about this, and the answer throws quite unexpected light on the political importance attained by clock-making at the end of the Northern Song. There is no doubt that the bronze components of Su's machinery were cast by imperial order in or just before AD 1090, after a period of trials superintended by one of the Hanlin academicians, Xu Jiang [Hsü Chiang]. This scholar, however, had comments to make which led to the construction of an entirely new piece of ancillary apparatus by Su Song, probably the earliest 'planetarium' which one could actually get inside. The passage is worth citing.

> In the third month of the fourth year of the Yuanyou reign-period (AD 1089) the wooden model (of Su Song's armillary clock-tower) was completed – previously there had been nothing like it. The Hanlin Academician Xu Jiang and others were ordered to examine it. (After due trials) . . . they said: 'Comparison has been made (of the armillary clock) with the dawns and dusks and the motions of the heavens, and there is perfect agreement, so the order has been given to cast it (i.e. the sphere and the globe) in bronze. Let it be called the Armillary Clock of the Yuanyou Reign-Period.'
>
> Some time afterwards (Xu) Jiang and others also said: 'Now what has always been called the armillary sphere has a spherical outer form, so that the (equatorial) constellation (-positions) in degree (-marks) can be distributed round it. Inside there are the concentric rings and the sighting-tube, which can be used for the observation of the heavens. (Thus we have here two separate instruments, the sphere and the globe, but they could be made into a single instrument). Now what has been erected (contains) the two

different things, the (turning) sphere and the (turning) globe, the former for observing the true numbers of the degrees of the heavenly motion, the latter placed within a closed room, performing the celestial revolution by itself, and checking what is observed by means of the armillary sphere. If these two instruments were to be combined into one, the globe would be (part of) the sphere, and together they would (show) correctly the heavenly motions, performing the function of both. (This also would be a contribution to the (astronomical) equipment of the present dynasty). So we request that an armillary sphere (of this kind) be commissioned.' It was decided to do this.

(Su) Song understood all these matters well because his own family had possessed a small model (globe or sphere). So he asked (Han) Gonglian to make the (necessary) elaborate calculations, and after several years the instrument was completed. It was larger than the height of a man, so that one could enter and sit inside it, its structure being like that of a lantern or a bird-cage (with bamboo ribs), and the walls (of silk and paper) pierced with holes according to the positions of the stars. It could be rotated by means of a wheel so that the star culminations for any particular time before dawn or after dusk could all be seen by looking through the holes. The astronomers and calendrical pundits all flocked to watch its operation, and were quite astonished at it, for nothing like it had previously been achieved.

Thus the auxiliary instrument made at Xu Jiang's suggestion was a planetarium in the sense that a human observer could enter and study an artificial sky, but not in the sense that the motions of the planets could be mechanically demonstrated. Though the language used resembles that applied so often to water-wheel drives, the word 'water' is missing in all sources, so the 'planetarium' may have been set by hand to whatever hour-angle desired; as in the elegant reconstruction recently proposed by Wang Zhenduo [Wang Chen-To] in 1962, who has the observer sitting in a chair which hangs from the main polar axis. Su Song doubtless met Xu Jiang to the extent of providing meridian and horizon rings, but the other great circles would be sufficiently marked on the skin of the world-ball. The year AD 1092 thus marked the peak of Su's practical achievements, though he had seven more years to live after finishing his book two years later.

Very soon, however, the masterpieces of Su Song and Han Gonglian were menaced by the controversies of the time. From about AD 1060 until the fall of the capital in 1126 the public life of the dynasty was torn by violent disputes between the party of the Conservatives and that of the Reformers. Here we cannot enlarge upon the issues which divided them though, as already mentioned, Su Song, while probably not an active member of the Conservatives, was identified with them through his friendships. So when in 1094 the Reformers came back into power there was talk of destroying his clock-tower. The idea was, of course, that each new reign-period must 'make all things new', and politicians of that time could hardly be expected to understand the slow growth of practical scientific knowledge. But in fact the clock remained untouched, and continued to tick over the minutes inexorably until those days of AD 1126 when the Jin [Chin] Tartars were at the gates.

In that fateful year the capital (Kaifeng) was twice under siege. In September it fell, and both emperors (Huizong [Hui Tsung] and Qinzong [Chhin Tsung]) were taken away captive to Beijing [Peking] in the north. For a while the princes and the remains of the court wandered about behind the lines, settling first in one place and then in another, till in 1129 it was decided to choose Hangzhou [Hangchow] as the new capital. This was what became the 'flower of cities all' that Marco Polo was to see 140 years later.

The blow at the technological supremacy of the Song was extremely severe. One of the most notable things about the sieges of Kaifeng had been the fact that in the periodical armistices the Jin Tartars exacted as tribute whole families of artisans and skilled workmen. They demanded from the city all sorts of craftsmen, and there is every reason for thinking that when the capital fell, all the clock-making millwrights and maintenance engineers followed the Jin power and migrated to the north. From comments in the *Jin Shi* [*Chin Shih*] (History of the Jin Dynasty) it seems likely that they accompanied the disassembled clock-tower itself.

The 15-tonne armillary sphere of Su Song eventually fell into the hands of the Mongols towards the beginning of their career of conquest. When Beijing became their capital in AD 1264 it was still available for the astronomical officials, but by then it had suffered from the ravages of time and could no longer conveniently be used.

Historians and sinologists have had much to say about the end of the Northern Song, but few if any have observed that it expired in a blaze of horological exuberance. The *Song Shu* [*Sung Shu*] (History of the Song Dynasty) tells us that in AD 1124 Wang Fu, (a minister of one of the princes) chanced to meet at the capital 'a wandering unworldly scholar' with the family name of Wang, who gave him a Daoist book which discussed the construction of astronomical instruments. The emperor agreed to authorise the construction of some models to test what the books said, and the chief instrument seems to have been a demonstrational armillary sphere or a celestial globe, with all the usual features and very elaborately graduated. But it was combined with complicated planetarium machinery which showed automatically not only the positions of the sun and moon but also the phases of the latter. The planets too were shown rising, culminating and setting, moving at varied speeds as they appeared to moved forwards and backwards across the sky. The text tells us that:

> A jade balancing mechanism (i.e. the escapement) is erected behind (literally, outside) a curtain, holding and resisting the main scoops. Water pours down, rotating the wheel. Lower, there is a cog-wheel with 43 (teeth). There are also hooks, pins and interlocking rods one holding another. Each (wheel) moves the next without reliance on any human force. The fastest wheel turns round each day through 292 teeth, the slowest only moves by 1 tooth in every 5 days. Such a great difference is there between the speed of the wheels, yet all of them depend on one single driving mechanism. In precision the engine can be compared with Nature itself (literally, the maker of all things). As for the rest, it is much the same as the apparatus made (long ago) by Yixing. But that old design employed mainly bronze and iron, which corroded and rusted so that the machine ceased to be able to move automatically. The modern plan substitutes hard wood for these parts, as beautiful as jade . . .

This description is one of the most interesting specifications for a Song astronomical clock that we have, tantalisingly abridged though it is. The use of hard wood is a particularly striking feature. Wang Fu proceeded to mention other technical aspects of the time-telling and striking components. He also proposed that a special bureau should be set up for constructing several machines of this kind, and also a special portable one to accompany imperial peregrinations. Eventually construction was ordered, but only two years afterwards there came the siege of the capital, and one must suppose that all the half-completed pieces, the designs and the artisans themselves were carried away to the north by the Jin Tartars. Wang Fu in any case did not live to know.

As to the Song dynasty, which moved south, astronomical and engineering science suffered such a blow that for a considerable time the needs of the imperial observatory could not be met. There was a demand for an astronomical clock, for in AD 1144 Su Song's son, Su Xi [Su Hsi] who had escaped to the south was called upon to search the family papers, but could produce no designs from which another machine could be made. Even the great Neo-Confucian philosopher Zhu Xi [Chu Hsi] (AD 1130 to 1200) interested himself in the clockwork drive and tried hard to reconstruct its mechanism, but without success.

The secret of the escapement being thus now temporarily lost, it appears that some Southern Song technicians returned to the simpler float principle of the anaphoric clock. However, Su Song's book was recovered in full and first printed in the south by Shi Yuanzhi [Shih Yuan-Chih] in 1172. Possibly Zhu Xi's agitation led to this result. In any case it is an interesting comment on the events in the south to find that an elaborate account of the works of Su Song's clock is contained in the history of the Jin Dynasty (*Jin Shu*) but not in the one for the Song (the *Song Shi*). Since these two dynastic histories were both edited by the same scholars, Toktaga the Mongol, and Ouyang Xuan [Ouyang Hsüan], in about 1350, the obvious inference is that the engineering description was conserved in the Northern Jin archives but not in those of the Southern Song.

   This observation restores us to the Mongol period, and to Guo Shoujing [Kuo Shou-Ching]. As early as 1262, before Beijing became Kublai Khan's capital, Guo had made for him 'a Precious Mountain Clock'. But in the *Yuan Shi* [*Yuan-Shih*] (History of the Yuan (Mongol) Dynasty) we have an elaborate description of the illuminated clock of the Da Ming [Ta Ming] Hall, made almost certainly by Guo, for it comes in the midst of a long account of his inventions, though not specifically ascribed to him. Horizontal wheels like Su Song's were used and the whole mechanism was enclosed in casing and driven by water. Though no water-wheel is mentioned, adequate power to operate the involved jack-work could not otherwise have been obtained. Indeed, there can be no room for doubt that Guo Shoujing was making clocks, more elaborate perhaps, but still in the same tradition as those of Yixing and Su Song. What is interesting, and new, is that the jack-work now completely dominates over the astronomical components. The clock, though not yet endowed with a name of its own, had almost wholly left the world of astronomy towards the end of the thirteenth century AD in China. This is a highly significant point, for about 1310 the same process had occurred in Europe. The exact date of the Da Ming illuminated clock is unsure, but it will not be far from 1276, the year when Guo Shoujing took in hand the restoration of the Beijing Observatory and its equipment with the best and newest instruments.

   These included one piece of apparatus, however, which, embodying a clock-work drive, presumably water-powered, perpetuated the age-old connection with astronomical uses. It has the added interest that it was seen and described by more than one of the Jesuit missionaries. This late representative of Zhang Heng was thus the first of its line to receive the scrutiny of observers from Europe. It was a bronze celestial globe 5.7 metres in circumference half sunk in a box-like casing within which were 'toothed wheels set in motion for turning the globe', so the *Yuan shi* [*Yuan Shih*] (History of the Yuan Dynasty) of the fourteenth century tells us. Ricci's own journal of the early spring of 1600 records that:

> . . . a globe, having all the parallels and meridians marked out
> degree by degree, and rather large in size, for three men with

outstretched arms could hardly have encircled it. It was set into a great cube of pure bronze which served as a pedestal for it, and in this box there was a little door through which one could enter to manipulate the works.

While a weight drive cannot be entirely excluded, this would have been so foreign to the Chinese tradition that a water-wheel is more likely.

We are now about to return to our original starting point, the Jesuits and their experiences in seventeenth-century China. But first we should take leave of the millennial indigenous tradition of water-powered clockwork. In order to do this, we must enter the private apartments of the imperial palace, where about the middle of the fourteenth century we find the last emperor of the Yuan dynasty, in extraordinary contrast to the hard-riding desert warriors who were his ancestors, busied – like Louis XIV – in making clocks himself. The *Yuan Shi* tells us that this engineering minded ruler, Shundi [Shun Ti]:

> . . . also constructed a Palace Clock 1.8 to 2 metres high and half
> as wide. A wooden casing hid many scoops which made the water
> circulate up and down within. On the casing was a 'Hall of Three
> Sages of the Western Paradise', at the side of which was a Jade
> Girl holding an indicator-rod for the (double-) hours and quarters.
> When the time arrived the figure rose up on a float, and to left
> and right appeared two genii in golden armour, one with a bell
> and the other with a gong. At night these jacks struck the night-
> watches automatically without the slightest mistake . . .

There was other elaborate jack-work, the ingenuity of which 'was beyond belief, and people said that nothing like it had ever been seen before.'

If this judgment is more flattering than accurate, one need not doubt that the imperial jack-work, though quite in the tradition of Yixing and Su Song, was impressive enough. More interesting and notable is the fact that this clock, though almost certainly without dial and pointer, had lost practically all trace of the original astronomical components. Yet in becoming a purely timekeeping machine it had not acquired any new name, and was merely called a 'leaker', like the simplest clepsydras of two thousand years before. It was of course far more, as we can deduce from the water-wheel escapement hinted at by the 'scoops'.

By curious coincidence, the clock was almost contemporary with the wonderful astronomical timepiece of Giovanni de Dondi in Italy (AD 1364). This translated astronomical data entirely into terms of dials and pointers, with some indications to the onlooker by inscribed links on endless chains, for rotating globes and spheres had been foreign to the European tradition.

Shundi's clock had already lost them, yet it was in direct line from the pre-clock of Zhang Heng; de Dondi's profited by the new verge-and-pallet escapement of its time, as also from the medieval computing devices like the equatorium, but it too was in direct line from the anaphoric dials and pointers of the Hellenistic age.

The final blow to this indigenous tradition might well be dated about AD 1368, when the new forces of the Ming dynasty captured Beijing and ended the Mongol domination. Yuan palaces and their contents were destroyed, for the Ming, like another revolution later, 'had no use for' clockmakers. No doubt the Chinese horological tradition had become smothered in its own jack-work, and was hopelessly identified with the 'conspicuous waste' of the Mongol court. But its death (if indeed it did quite die at that time) was a circumstance of peculiar historical importance, for it meant that when the Jesuits arrived two hundred and fifty years later there was extremely little to show them that mechanical clocks had ever been known in China.

Though as we have seen, the Jesuits believed the mechanical clocks they brought with them were something entirely new to the Chinese, nevertheless they spoke in rather mysterious terms of timekeeping methods previously in use in China. An interesting passage is contained in the admirable description of China which Ricci prefixed to his memoirs, and this was enlarged by Trigault as follows:

> They (the Chinese) have very few instruments for measuring time, and those which they do have measure it either by water or fire. Those which use water are large clepsydras, and those which use fire are made of certain odiferous ash very like that tinder which is used for the touch-wood or slow match of guns and cannon among us. A few other instruments are made with wheels rotated by sand as if by water – but all are mere shadows of our mechanisms, and generally most faulty in time-keeping.

We are thus obliged to believe that still at the time of Ricci and Trigault some remains were left of the old Chinese driving-wheel clocks. The use of sand seems particularly strange, for the hour-glass is considered to have been an introduction from Europe then of very recent date. Nevertheless sand there was and what happened in the Ming was unravelled by Yabuuchi Kiyoshu and Liu Xianzhou [Liu Hsien-Chou] in the 1950s. We have just seen how the monumental striking water-clock of the last Yuan emperor perished, but if the Ming emperor Taizu [Thai Tsu] felt compelled to frown upon anything which savoured of the palatial luxury of alien rulers, he seems to have given positive encouragement to a form of wheel-clock which could be easily made in all provinces and prefectures. The texts which tell us of this new horological design are also in the first of the astronomical chapters of the *Ming Shi* [*Ming Chih*] (History of the Ming Dynasty), but they concern a much later date, and the matter arises incidentally. The Chinese astronomers who were the friends and collaborators of the Jesuits are discussing the making of new equipment for the imperial observatories.

> In the following year (AD 1635) (Li) Tianjing [Thien-Ching] suggested that sand-clocks should be made. At the beginning of the Ming (i.e. about AD 1360 to 1380) Zhan Xiyuan [Chan Hsi-Yuan], finding that in bitter winters the water froze and could not flow, replaced it by sand. But this ran through too fast to agree with the heavenly revolution, so to the (main driving) wheel with scoops he added four wheels each having thirty six teeth. Later on, Zhou Shuxue [Chou Shu-Hsüeh] criticised this design because the orifice was too small so that the sand-grains were liable to block it up, and therefore changed the system to one of six wheels (in all), the five wheels each having thirty teeth, at the same time slightly enlarging the orifice. Then the rotation of the machine really agreed with the movement of the heavens. What (Li) Tianjing now petitioned for was surely this design deriving from (Zhan and Zhou).

In this way we come across a new name, hitherto unknown, but of much importance for our story, Zhan Xiyuan from AD 1370. Zhou Shuxue we have met before; he flourished between AD 1530 and 1558 leaving a well-deserved reputation as a mathematician, astronomer and cartographer. The date of Zhan Xiyuan is clearly confirmed by the fact that the great historian Song Lian [Sung Lien], chief editor of the *Yuan Shi*, wrote an interesting essay on the 'Five-Wheeled Sand-Clock and its Inscription', prepared by him for one of these instruments before his death in 1381. The account of the mechanism given in this text is so complete that Liu Xianzhou had no difficulty in making the working drawing shown in Fig. 378. Two things are very clear, first that this type of sand-clock had a scoop-wheel very similar to that in the great water-wheel clocks of Su Song and others, as well as a certain amount of jack-work as they did; secondly that it had a stationary dial-face over which a pointer circulated. The appearance of this feature, so characteristic of all subsequent clocks, in late fourteenth century China, is quite curious, since this was just about the time when the dial-face was becoming standardised in Europe too. Perhaps the simplest hypothesis is that both derived independently from the dial of the anaphoric clock, which as we have seen (p. 238), seems to have existed during the earlier Middle Ages in China as well as in the West. Equally important is the evidence of the divorce, now complete,

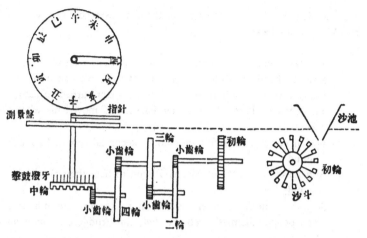

Fig. 378. Reconstruction by Liu Xianzhou of the mechanism of one of the sand-driven wheel clocks of Zhan Xiyuan (*c.* AD 1370). Its five wheels comprised the driving wheel with scoops below the feed, three large gear wheels and one 'middle wheel' fitted with audible signal trip-lugs and borne on the shaft of the pointer which made the rounds of the dial face. On this the markings for the twelve double-hours can be seen.

both in China and Europe, between the astronomical functions and the purely time-telling functions of clockwork and proto-clockwork.

The most difficult thing to decide about the sand-clocks of the Ming is whether or not they worked with a linkage escapement like the earlier monumental water-wheel clocks from Yixing to the Yuan emperor Toghan Timur (reigned AD 1333 to 1367). It would have been natural for them to have possessed it, but there is no justification in Song Lian's text for supposing that they did, and it may be, therefore, that Zhan Xiyuan was more original than Song Lian realised. For he was, perhaps, the inventor of a type of clock with reduction gearing. Computation shows that whether in its original or its modified form, each scoop took only a matter of seconds to fill, and with such an extensive train of gears there could have been no tendency for the driving-wheel to run wild. Of course, with the level of technique available in the fourteenth century, or even the sixteenth, the resulting accuracy may well have deserved the strictures it received from the Jesuits, and no doubt the advent of weight-and spring-driven clocks was a very great gain; yet the work of Zhan Xiyuan retains all its interest. Reduction gearing had been known in principle to the Alexandrians, and for timekeeping machinery it was of course a commonplace in fourteenth century Europe, but the thought of relying completely upon it for slow motion relayed from a prime mover does not seem to have occurred to contemporary Europeans.

Sung Lian even wrote a poem about them in his account of the history of the Ming sand-clocks:

> The 'Hoisters of the Water-Pots' were ancient men of State,
> And as of old the time they told by Water's constant rate,
> But wintry ice spoilt their device – the water turned to land –
> Till Zhan by Earth did conquer Earth, and moved his wheels with
>   sand;
> Which being neither firm nor flood, keeps faith with Heaven's
>   round
> And makes the jacks of Master Zheng to beat their rhythmic
>   sound.
> So now Stone flows where Water can't, ignoring fire and frost –
> Good people all, mark well the dial, and grudge each moment lost.

And indeed the sand-clocks of the Ming are worth our careful notice too. Their period of dominance was comparatively so recent that the discovery of an actual specimen in the excavation of a Ming tomb is certainly possible.

A late appearance of solid particles fulfilling the functions of water in the 'mill-wheel' type of clock brings us to the point where the two traditions, Chinese and Western, joined in a single instrument. By the second decade after the death of Ricci collaboration between Jesuit missionaries and Chinese scholars had become a well-established tradition, and it is to the work of one such pair that we owe the earliest Chinese description of the verge-and-foliot escapement. But in this clock the two traditions fused, for while a mechanism of European type kept equal double-hour time in the front of the cabinet, a scoop-wheel system of Chinese type told the night-watches at the back.

It seems probable that the bucket-wheel sand-driven clocks did not die out until the close of the seventeenth century AD. Some time about 1660 a magistrate in Yunnan, Ji Tanran [Chi Than-Jan], constructed what was almost certainly an instrument of this kind for a Buddhist temple. Built in the form of a pagoda, it possessed a double-hour drum carrying placards to indicate the time, a wooden figure which appeared every quarter-hour; bells were also sounded. We do not know for certain the motive power used; the drum carrying the placards was horizontal like the jack-wheels of Su Song, but perhaps it operated by reduction gearing. Certainly it soon went wrong and ceased working altogether, and we have the impression of attending the death-bed of a great tradition.

In this section we have covered six hundred years. We saw how the remains of Su Song's great clock came into the possession of the Mongolian rulers and their brilliant men of science such as Guo Shoujing in the thirteenth century AD, some of whose own water-powered astronomical instruments were handed down to be studied by Jesuits at the end of the sixteenth century. In one of the Yuan emperors, Toghan Timur, we found a clock-maker of note, apparently the last in China to use water as the motive force; for from the beginning of the Ming (about AD 1370) Zhan Xiyuan and others substituted tanks of sand. This method survived well into the Jesuit period, but by the end of the seventeenth century, sand and scoop-wheels had given place to the standard forms of Europe, first the falling weight and then the spring. We have come a long way, but something yet remains to be said about developments which occurred during the Ming dynasty, from the fifteenth century AD onwards.

KOREAN ORRERIES, ASIATICK SING-SONGS, AND THE MECHANISATION OF MT MERU

When the Ming artisans broke down the carved work of the Yuan palaces, some of Shundi's skilled workmen may have got away in time. We can hazard a shrewd guess as to where they went. For a few decades later the court of the Yi kingdom of Chosŏn at Seoul in Korea was busied with exactly the same kind of activity.

The fourth king of this dynasty, Sejong (reigned AD 1419 to 1450) maintained a court not unworthy to be compared with those of Caliph al-Ma'mūn or Alfonso X, king of Castile. An enlightened and scholarly ruler, he was the prime inventor of the Korean alphabet in 1446, but was also fond of science and scientific instruments and supervised himself the complete re-equipment over seven years of the capital's astronomical observatory. This occurred from 1432 onwards, and many instruments were constructed. Unlike a noted chiming clepsydra at the capital constructed in 1434 in what seems to have been Arabic style, two of the new pieces of equipment were mechanised by water-wheel link-escapement clockwork. One was a one-metre diameter celestial globe made of lacquered cloth with a threaded model of the sun, the other a puppet-clock, and descriptions show that both followed the Yixing, Su Song and Shundi tradition.

A new chapter opened when in AD 1657 King Hyojong ordered an astronomical clock to be made. This had a water-wheel linkwork escapement. Two others were then made, but only one used a water-wheel drive, the other having 'Western gear-wheels' and seems to have been weight- or spring-driven. Thus we find both a fidelity to ancient Chinese tradition and yet another turning point, for just as the water-wheel system replaced the anaphoric striking clepsydra in 1438, so 1664 sees a change from the water-wheel system to modern clockwork.

In the seventeenth and eighteenth centuries AD, both before and after the Jesuit dominance in Chinese clockmaking had given way to the making of 'sing-songs' (as mechanical toys with fanciful jack-work became called), the horological traditions of China had a much greater effect on European design than has generally been realised. Features emanating from the ancient tradition of Yixing and Su Song were incorporated in products destined for the 'Asiatick trade', and in Europe itself complex dials and elaborate jack-work became popular in a context of 'Chinoiserie'. Incidentally, it was also in the seventeenth century in an entertaining book on commercial life, the *Nippon Eitai-gura* (Japanese Family Storehouse; or, the Millionaire's Gospel Modernised) of 1685 by Ihara Saikaku that the Japanese acknowledged that mechanical clocks had been invented in China.

CLOCKWORK AND INTER-CULTURAL RELATIONS

A provisional scheme of inter-relations is shown in Table 50. From this it is shown that the factors leading to the first escapement clock in China were a succession of 'pre-clocks' starting with Zhang Heng about AD 125, then leading on through others to Yixing (AD 725). As for gearing we have seen evidence that much experimentation with it was carried out through the centuries by Chinese mechanics.

The factors leading to the first escapement clock in Europe (*c.* AD 1300) were the weight-drive descended no doubt from the floats of the Hellenistic anaphoric clocks and mechanical puppet theatres, and it was certainly known in its free form in the thirteenth century AD, as the Moorish drum water-clock of Alphonsine times demonstrates. The use of gearing descended from remote antiquity, for even if we know little of the nature of a planetarium ascribed by Cicero and Ovid to Archimedes (*c.* 250 BC), the Anti-Kythera 'computing machine' from ancient Greece, with its elaborate gear-wheels, remains to show the extraordinary attainment which Hellenistic (first century BC) technique could reach. Then in the Arab realm there was the application of gearing to the computing astrolabe. The clock dial may be considered ultimately a derivative of the revolving dial of the anaphoric clock. Only the escapement itself had to be supplied by the unknown inventors of AD 1300. The verge-and-foliot itself probably derived from the radial bob flywheel, familiar since the early days of the Graeco-Roman screw-presses. But just how original was the basic idea? The preceding six centuries of Chinese escapements suggest that at least some stimulus diffused from east to west.

To gain a little light on this transmission, if such it was, we must concentrate attention on the years between about AD 1000 and 1300 and see whether any help can be found from the Islamic and Indian culture-areas. For example, writing on striking clepsydras in 1206, Ismāʾīl ibn al-Razzāz al-Jazarī refers to a tipping bucket attached to a hinged ratchet which pushes round a gear-wheel by one tooth each time a bucket fills with water and comes to the emptying point. Now al-Jazarī also had water-wheels in his clocks, a fact which can hardly be without significance in relation to earlier Chinese practice. Moreover, there is nothing in the evidence so far available which suggests any Arabic influence on the Chinese developments. On the other hand, the Arabic material does indicate the passage westwards of certain Chinese elements. Turning to India, we find that the mathematical and astronomical treatise *Siddhān Śiromaṇi* by Bhāskara written about AD 1150 discusses perpetual-motion machines using scoops filled with mercury, and also refers to an armillary sphere for demonstration purposes sunk in casing to represent the horizon. The text also says that 'by the application of water ascertainment is made of the revolution of time', while it is clear also that mercury or oil were also used.

For Hindus as well as Daoists, the universe itself was a perpetual-motion machine, and India thus provides an undoubted association between armillary spheres, time-measuring water-wheels and devices for perpetual motion. Even Zhang Sixun's mercury drive seems to be present.

The perpetual-motion machine makes its appearance in Europe in the notebooks of Villard de Honnecourt (AD 1237), but much more significantly in the great work on the magnet by Petrus Peregrinus (AD 1267). Again, as we have already seen (this abridgement, volume 3, p. 9), the first knowledge of the magnetic compass in Europe may be dated close to AD 1190. Since there is no remaining doubt that it travelled thither from the Chinese culture-area, the most obvious conclusion is that the carriers of the transmission, whoever they were, spoke not only of the magnet but also of armillary spheres perpetually in motion urged by their tireless wheels.

Though it is customary in modern science to despise the seekers after perpetual motion, in its correct historical perspective the idea had a stimulating value for enquiry. As early as AD 1235 William of Auvergne, then Bishop of Paris, used magnetism to account for the perpetual motion of the celestial spheres. Then in 1514 Amadeus Meygret maintained that since it would be turning under the direct influence of the heavens, the usual objections to perpetual motion could not hold. Indeed if one thinks of this idea as an attempt to capture the daily rotation of the heavens as a source of terrestrial energy, it seems eminently reasonable, though based on the false assumptions of its time. At the end of the sixteenth century William Gilbert, who thought of the earth as a giant magnet, used the idea to solve some of the objections to the sun-centred theory of Copernicus, which he himself defended. Even more important, the example of the field of magnetic attraction invisibly extended in space led directly, through Gilbert and Johannes Kepler, to Newton's concept of universal gravitation. Thus the adopted Indian belief in the possibility of perpetual motion, allied with the transmitted Chinese

Table 50. *Roles of different culture-areas in the development of mechanical clockwork*

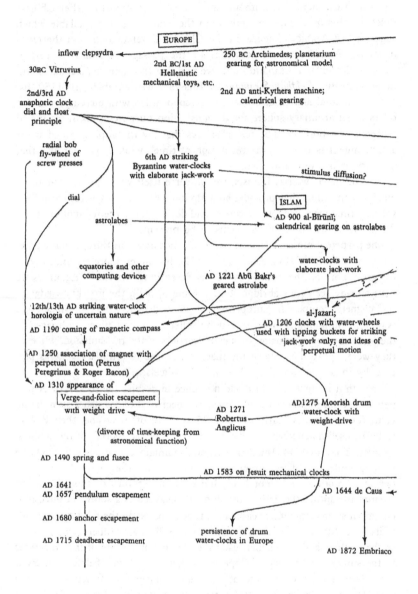

Table 50. *contd*

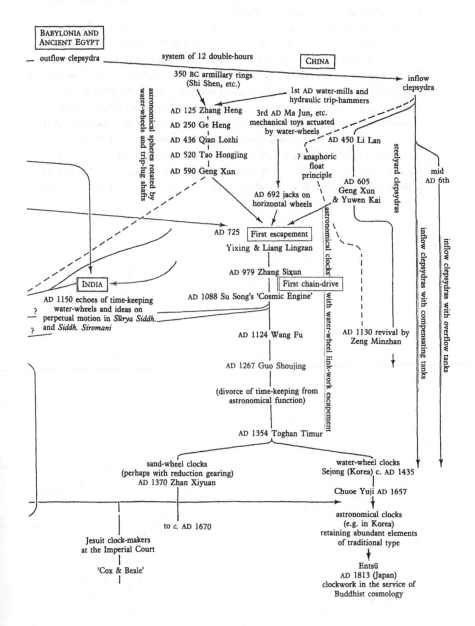

knowledge of magnetic polarity, deeply influenced modern scientific thought at one of its most crucial early stages.

Truly they influenced technology too. As already mentioned at the beginning of this chapter, Nicole Oresme (died AD 1382) was the first to use the metaphor of the universe as a vast clock set in motion with all harmony by God. Yet even by the middle of the thirteenth century, many active thinkers stimulated not only by the technological successes of recent generations but also led on by the idea of perpetual motion, began to generalise the concept of mechanical power. They came to think of the cosmos as a vast reservoir of energy to be tapped for human use. So it is clear also that without the earlier speculations of the Chinese and Indian naturalists, there might well never have been such a fantastic concept nor the power. As the *Guan Zi* [*Kuan Tzu*] (The Book of Master Kuan) said in the fourth century BC, 'The sage commands Nature and is not commanded by Nature.'

6

Vertical and horizontal mountings, windmills and aeronautics

In the foregoing chapters we have become very conscious of the distinction necessary in the history of technology between wheels mounted vertically and wheels mounted horizontally. The vertical wheel often necessitated right-angle gearing, and the horizontal required a very substantial lower bearing since it had to take all the weight. They have a different distribution in time and space. In what we must now examine, we shall see that horizontal wheels are again characteristic of China and indeed of all Central Asia and Iran, both as regards certain wheels turned by hand in libraries, and as regards wheels rotated by a new source of power, the wind; while conversely the vertical wheel continued to exert its influence in the West.

The revolving bookcase is not perhaps a very inspiring subject from a modern technological point of view, for it led to nothing further, though in its heyday, as one of the sacred or numinous aspects of great Buddhist temples, it deeply affected many Chinese Confucian scholars not easily led astray by their emotions. Its interest for us lies rather in the fact that it illustrates perfectly the preference of Western engineers for vertical, and of Chinese engineers for horizontal mountings. But it is also interesting as fur-nishing another case, parallel with the chain-excavator referred to above (chapter 3, p. 137), in which the Jesuits introduced to their Chinese friends, as something new, a device which had been known in China nearly a thousand years previously.

The *Qi Qi Tu Shuo* [*Chhi Chhi Thu Shuo*] (Diagrams and Explanations of Wonderful Machines) of AD 1627 pictured a revolving bookcase (Fig. 379) by the aid of which a scholar could consult a number of books without moving from his seat; simple gearing ensured that the books should remain right side up as the structure revolved. This was a copy of a design which appeared first in 1588 in a work of the Italian engineer Agostino Ramelli. Certainly this was a transmission from west to east.

However, prototypes of the revolving bookstand existed in China from the sixth century AD, for it had then become customary to erect large bookcases

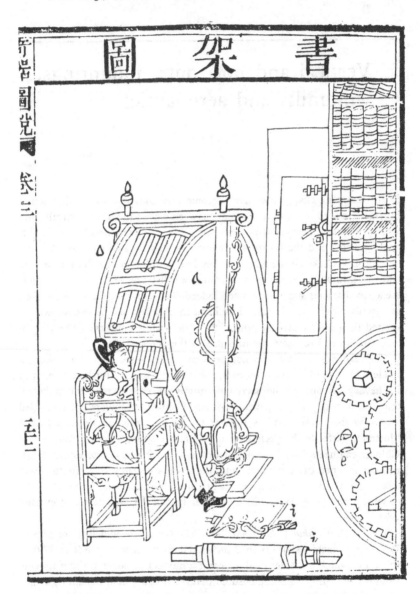

Fig. 379. Revolving bookcase figured in the *Diagrams and Explanations of Wonderful Machines* of AD 1627. Although the books in it have vertical rulings for Chinese type, those on the shelves are bound in European style.

in Buddhist temples so that the whole of the scriptures could be contained in them. During the course of centuries whatever practical use they may once have had faded into the background, and their chief significance became ritual, persons wishing to acquire merit being able to turn the whole structure round and thus perform the symbolic act of 'turning the Wheel of the Law'.

The traditional inventor is the scholar Fu Xi [Fu Hsi] in AD 544, and the bookcases certainly date from his century, even though literary references to them only go back to 823. Evidence becomes really abundant in the eleventh century, perhaps because of the printing of the Buddhist Canon shortly before (971 to 983). Ye Mengde [Yeh Mêng-Tê] wrote about AD 1100: 'When I was young I saw several four-sided revolving repositories. Recently in all centres large and small, . . . in six or seven out of ten temples, one can hear the sounds of the wheels of the revolving cases turning.' In the Song nine examples were particularly famous, and then after a break during the Yuan, we have fourteen descriptions of the building of such structures under the Ming, the latest in 1650 – significantly some time after the designs in the *Qi Qi Tu Shuo*, but completely disregarding them.

These Buddhist revolving repositories were all as horizontal, technically, as the Chinese horizontal water-wheels. One that still exists at the Longxing [Lung-Hsing] Temple at Zhengding [Chêngting] in Hebei [Hopei] has been examined and is shown in Fig. 380, and probably dates only from the Song, though the temple itself goes back to AD 586. From a great Song treatise on architecture, the *Ying Zao Fa Shi* [*Ying Tsao Fa Shih*] (Treatise on Architectural Methods) (AD 1103) the 'Turning Treasury of the Sūtras' is discussed in two places. Its height was to be 6 metres and its diameter 4.9 metres, and this is just about the size of the existing specimens, but others were certainly very much larger, approximating to the 21.3 metres of the prayer-cylinders at the Yong He Gong [Yung Ho Kung] in Beijing [Peking]. Some took the strength of ten men to revolve.

For the purposes of the Buddhists, no gear wheels were necessary, yet it does not follow that gearing was never used. There may be an obscure reference to this in Yang Shen's sixteenth-century AD *Dan Qian Zong Lu* [*Tan Chhien Tsung Lu*] (Red Lead Record), where 'male and female' wheels revolving in opposite directions are referred to; and the repository of the Kaifu [Khai-Fu] Temple near Changsha [Chhangsha] had in AD 1119 five wheels which all turned together. It has been suggested that the original reason for the invention of the revolving bookcases may well have been connected with the great burden of translation work assumed by Chinese Buddhists during the early centuries of our era. Wheel or cylinder libraries never arose in India or Central Asia, nor among Chinese Confucians or Daoists.

Repositories for the scriptures were not the only pieces of equipment which revolved horizontally in medieval China. There are hints that certain 'bedside

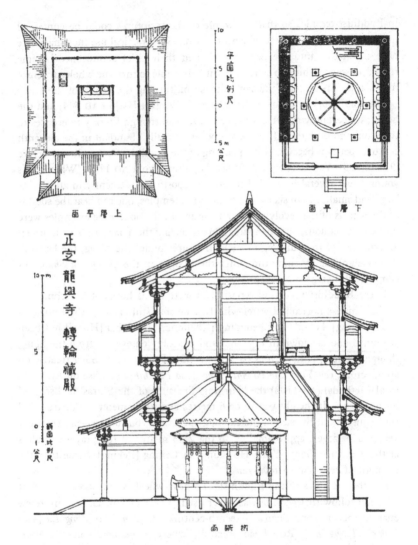

Fig. 380. Scale drawings of the Repository Hall of the Longxing Temple at Zhengding in Hebei. Above, left, upper storey plan, right, ground-floor plan; below, section in elevation.

bookcases' rotated as early as the third century AD, while swivelling chairs first appear about AD 345. Certain, too, was the wheel-shaped case for the systematic storage of a fount of movable type, introduced by Bi Sheng [Pi Sheng] about AD 1045.

THE WINDMILL IN EAST AND WEST

The earliest records in Europe of the windmill date from AD 1180. From this time onwards windmills spread rapidly, coming into use throughout Western countries during the thirteenth century. No illustration, however, survives from before about AD 1270, the date of the so-called 'Windmill Psalter', probably written at Canterbury. After 1349 they become numerous.

The Western vertically-mounted windmill was from the beginning a kind of reversed or ex-aerial propeller, that is to say wheels or windmill sails so designed as to make use of the force of the winds for doing work. It was no doubt essentially a purely practical development, derived from the Archimedean screw and not from the water-mill of Vitruvius. It was thus deeply Western, but it involved one new mechanical problem, the orientation of the main driving-shaft (the 'windshaft') so as to present the sails (or 'wheel') at right angles to the direction of the wind. Two distinct types appeared. There were smaller mill-housings revolving round a central post or pivot permanently fixed on (or in) the ground; these were the 'post-mills'. Then there were the larger mills consisting of a tower of brick or masonry roofed by a movable cap carrying the sails and the shaft – the 'tower-mills'. All the earlier mills were of the post type, supported by four diagonal slanting legs (the 'quarterbars').

The history of windmills really begins in Islamic culture, and in Iran. In a story due to 'Alī al-Ṭabarī (c. AD 850) and others, the second orthodox Caliph, 'Umar ibn al-Khaṭṭāb, was murdered in AD 644 by a captured Persian technician Abū Lu'lu'a, who claimed to be able to construct mills driven by the power of the wind, but was bitter about the high taxes to which he had been subjected. More certain, perhaps, is the mention of windmills in the works of the Banū Mūsā brothers (AD 850 to 870), while a century later several reliable authors are speaking of the remarkable windmills of Seistan, an arid sandy region in southeast Iran, renowned for the continuous blowing of high winds. A very detailed description of them occurs in the *Nukhbat al-Dahr* (Cosmography) of Abū 'Abdallāh al-Anṣārī al-Ṣūfī al-Dimashqī about AD 1300. From this it is clear (Fig. 381) that the Iranian windmills were horizontal in type, and enclosed in shield-walls so that the wind entered only on one side, in turbine-fashion. They still grind on today, though while retaining their shield-walls, their wheels have become greatly broadened so as to form tall upstanding rotors, and the millstones have been placed beneath.

 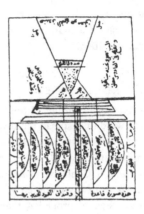

Fig. 381. Diagrams of windmills from the cosmographical treatise of al-Dimashqī *c.* AD 1300. The horizontal rotors have their vanes below, the millstones being in an upper storey.

The first European to see windmills in China was Jan Nieuhoff, who met with them at Baoying [Paoying] in Jiangsu [Chiangsu] when journeying north along the Grand Canal in AD 1656. Such windmills as he described are still extensively used all along the eastern coast of China north of the Yangzi [Yangtze], and particularly in the region near Tianjin [Tientin], mainly as prime movers operating square-pallet chain-pumps by right-angle gearing in the numerous places where salt is made from sea-water. This construction is of particular interest, for the vanes or surfaces taking the wind-pressure do not radiate from the central axis, but are in fact true slatted sails as used in Chinese junks, (see this abridgement, volume 3, pp. 181 ff), mounted on eight masts forming the rim of a skeleton drum. (Fig. 382). The scholar Chen Li who has closely studied them quotes a local riddle which helps us to understand the construction:

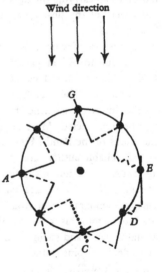

Who is the great general with the eight faces,
Strong in the teeth of the wild winds?
He has eight masts that follow the wind and turn.
Wearing a hat at the top, and standing on a needle below,
His two ends can revolve at your wish,
And make waters come or go wherever you like.

The 'hat' is the upper bearing of the central axle, and the 'needle' is the pin
or gudgeon on which it revolves below. But the ingenuity of the whole device
is that it dispenses entirely with the shield-walls used in Iran.

Unfortunately, there are very few literary references in Chinese to this or
any other windmill. The only really important reference so far relates to the
borders of Xinjiang [Sinkiang]; it occurs in the *Shu Chai Lao Xue Cong Tan*
[*Shu Chai Lao Hsüeh Tshung Than*] (Collected Talks of the Learned Old Man
of the Shu Studio), a book of the Yuan or late Song time by Sheng Ruozi
[Shêng Jo-Tzu]. He says:

In the collection of the private works of the 'Placid Retired
Scholar' there are ten poems on Hezhong Fu [Ho-chung Fu].

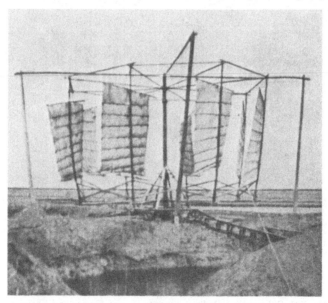

Fig. 382. Typical Chinese horizontal windmill working a pump in the
salterns at Dagu, Hebei. The fore-and-aft matt-and-batten type sails pivot at
certain points in the cycle and oppose no resistance as they come back into
the eye of the wind. (See also diagram on opposite page.)

One of these describes the scenery of that place ... and says
that 'the stored wheat is milled by the rushing wind ... The
westerners there use windmills just as the people of the south use
water-mills ...

When we realise that the 'Placid Retired Scholar' was Yelu Chucai [Yehlü
Chhu-Tshai], the great Jin and Yuan statesman and patron of astronomers
and engineers, and that this passage must therefore refer to the year AD 1219,
when he visited Turkestan, and also that Hezhong Fu was the place we know
as Samarkand, there can remain no doubt that the northern Chinese must have
been acquainted with the Persian windmill during the thirteenth century. The
Western Liao State was thus almost certainly the focus of this transmission.
The windmills of the Persian culture-area were described more fully some
two hundred years later by other Chinese visitors.

   The most probable supposition, therefore, seems to be that the horizontal
windmill was introduced into China either by Central Asian, Qidan [Chhi-tan]

Fig. 383. Oblique-axis windmill characteristic of Jiangsu and Zhejiang.
Although these windmills are fitted with typically Chinese mat-and-batten
type sails, and work, as here, pumps for irrigation, they probably derive
from Dutch originals of the seventeenth century AD with Western sails
which turn Archimedean screws fitted directly on the windshaft.

or Arab–Persian merchants overland, or by Arab–Indian sailors or merchants through the ports, some time during the Song or Yuan periods. This transmission may only have been the message that wind-power could be used with a horizontal rotor, whereupon Chinese nautical technicians proceeded to construct for their friends in the salt industry the 'fore-and-aft' windmill as we now have it. This explains at any rate why the distribution remained coastal; inland, away from the great rivers, there were no skilled sail-makers.

Jesuit influence may have left its mark in China in another way, however. In certain of the eastern Chinese provinces and particularly between Shanghai and Hangzhou [Hangchow], wind-power is harnessed for water-raising by means of a curiously constructed windmill the axis of which is set neither vertically nor horizontally but obliquely (Fig. 383). Now from the sixteenth century AD onwards in Holland very similar small windmills have been in use, their axles being continuous with the inclined shafts of Archimedean screws. Most probably, therefore, the oblique windmill was introduced into China in the seventeenth century as part of a compact piece of equipment which included the Archimedean screw.

To sum up, the Chinese windmill is a characteristic contribution of its own, derived, we may certainly say, from the typically Asian horizontal windmills of Iran, but embodying devices borrowed from nautical technique so ingeniously as to make it almost a new invention. The European vertical windmill, on the other hand, equally original, though probably also due to a stimulus from Iran, seems more likely to have derived from the Archimedean screw and the vertical as well as the horizontal water-wheel. The origins of the Persian horizontal windmill must remain at present undetermined, but they seem likely to have had as much to do with the Mongol–Tibetan wind-driven prayer-mechanisms as with the horizontal water-wheel or the toys of ancient Alexandria.

THE PREHISTORY OF AERONAUTICAL ENGINEERING

In the foregoing section we have been concerned with wheels or rotors which, according to Dr Needham's terminology, might be called 'ex-aerial' (see p. 271 above). In earlier sections we also saw that 'ad-aerial' wheels or fans were known to ancient and medieval China whether for use in winnowing, or for cooling palace halls in summer weather. Although these employments did not involve the motion of any vehicle, which is one of the greatest uses of ad-aerial rotors today, we shall shortly see not only that some precursors of the aircraft propeller existed in China, but that one of them played a cardinal part in the development of modern aerodynamic thinking. More, the rise of this new science in the nineteenth century depended fundamentally upon the study of an apparatus which had not been known in Europe before the sixteenth, and which was partly Chinese in origin, namely the kite. The kite's stretched fabric is of course not a shaped aerofoil like the wing of an aircraft, but the most essential difference between them is the fact that the lift of the kite is provided by a fortuitous air-stream of the wind, while that of an aircraft's wing is made mechanically by its propellers. Pilots of today who call aircraft 'kites' perhaps hardly realise how fitting historically is this slang term.

'Lie Zi [Lieh Tzu] could ride upon the wind. Cool and skillfully sailing, he would go on for fifteen days before returning . . .' These words from the section on Daoism in the *Zhuang Zi* [*Chuang Tzu*] (The Book of Master Zhuang) of 290 BC give us, indeed, just the proper starting point for the present review. For though Daoists elevated the idea of 'making excursion on the winds' and 'riding on the immensity of the universe' to philosophical heights, the idea itself had grown out of that primitive Asian shamanism, which was one of the roots of their school.

Legendary material

Legends of self-propelled aerial-cars, as opposed to flying vehicles drawn by winged animals and to unassisted personal flight in the style of Daedalus and Icarus, go back a long way in China, where they were associated with the mythical person or people called Jigong [Chi-Kung]. Though mention of them appears in the Early Han, there is no mention of their aircraft. This appears suddenly in the works of two third-century AD contemporaries, Zhang Hua [Chang Hua] and Huangfu Mi. The former, in his *Bo Wu Zhi* [*Po Wu Chih*] (Record of the Investigation of Things), says:

> The Jigong people . . . could also make aerial carriages which, with a fair wind, travelled great distances. In the time of the emperor Tang (the legendary founder of the Shang dynasty, and therefore second millennium BC), a westerly wind carried such a car as far as Yuzhou [Yüchow], whereupon Tang had the car taken to pieces, not wishing his own people to see it. Ten years later there came an easterly wind of (sufficient strength), and then the car was reassembled and the visitors were sent back to their own country, which lies 40,000 *li* (20,000 km) beyond the Jade Gate.

Exactly the same story occurs in a book by Huangfu Mi, and is echoed time after time, for example in the fifth and sixth centuries.

There is a tradition of pictures associated with the story, the oldest coming from *Yi Yu Tu Zhi* [*I Yü Thu Chih*] (Illustrated Record of Strange Countries), completed some time after AD 1392 and printed in 1489. It shows a rectangular chariot with two occupants and one curious wheel which appears to be toothed. A variant appears in editions of the *Shan Hai Jing* [*Shan Hai Ching*] (Classic of the Mountains and Rivers) of Zhou and Han times, but again this

Fig. 384. A pictorial version of the aerial car of the Jigong people, from the *Classic of the Mountains and Rivers* (text of the second century BC or earlier, with seventeenth-century AD commentary). The inscription follows the original text fairly closely at first, saying: 'The people of the Jigong country have each one arm and three eyes, and they are partly male and party female. They are able to construct flying carriages which can follow the wind and travel great distances . . . by studying the winds they created and built flying wheels with which they can ride along the paths of whirlwinds. They visited us in the time of the emperor Tang.'

shows an attempt to depict screw-bladed rotors or propellers (Fig. 384). Further on we shall produce evidence that the possibilities of the helicopter top or 'Chinese top' were appreciated as early as the fourth century AD, and it may be, therefore, that some of the medieval artists who depicted the car of Jigong were able to imagine the applicability of such rotors to horizontal motion. Flying cars drawn by birds, griffons and dragons were a separate tradition, which started in the Han, though its origins are probably Babylonian.

Thaumaturgical artisans

From the writers and artists we must now pass to the miracle workers or thaumaturgical artisans. In the end, someone actually does something. The invention of a wooden kite (*mu yuan*, 木鷂) is ascribed in ancient texts to Mo Di [Mo Ti] (died 380 BC), the founder of the Mohist school (volume 1, chapter 9 of this abridgement), and to his contemporary Gongshu Pan [Kungshu Phan], the famous engineer of the State of Lu. Whether it was in the shape of a bird is not clear. The character *yuan* continued to mean the bird which we call a kite (*Milvus lineatus* and related species), and when applied to the flying device was usually qualified by the adjectival word *zhi* [*chih*] (鷂). The

Han Fei Zi [*Han Fei Tzu*] (The Book of Master Han Fei), written about 255 BC, says: 'Mo Zi made a wooden kite which took three years to complete. It could indeed fly, but after one day's trial it was wrecked.' The *Mo Zi* [*Mo Tzu*] book, fourth century BC, has a closely similar statement, and tells us that Gongshu Pan constructed a bird from bamboo and wood, which stayed aloft for three days without coming down. In later times everybody knew these stories, one of them saying that Gongshu Pan's kite swayed and somersaulted, another that the devices of Mo Di and Gongshu Pan were kites, such as were flown by Song children. One tradition even goes so far as to say that Gongshu Pan flew wooden man-lifting kites over the city of Song during a siege, either for observation or as vantage-points for archers. If this should be considered unlikely for the fourth century BC, we shall nevertheless see in a moment that the military use of kites goes back a long way in Chinese history.

About AD 43, the great sceptic Wang Chong [Wang Chhung] sought to discredit the traditions about Mo Di and Gongshu Pan, though he did not disbelieve in the possibility of artificial flight. Certainly another attempt seems to have been made by his younger contemporary, the great astronomer and engineer Zhang Heng [Chang Hêng]. The main information is in the *Wen Shi Zhuan* [*Wên Shih Chuan*] (Records of the Scholars) by Zhang Yin [Chang Yin] published during the Jin dynasty. Here it is said that a wooden bird was made, with wings and pinions, having in its belly a mechanism which enabled it to fly several *li*. Dr Needham thinks that the devices of Mo Di and Gongshu Pan were kites, probably shaped roughly like birds, but that of Zhang Heng could have involved the air-screw of the helicopter top, though the only motive power available to him for such a purpose would have been springs. He is also of the opinion that the statement of the distance flown should not be taken too seriously. As for Zhang Heng, there are references to the machine in his own writings. In his *Ying Xian* [*Ying Hsien*] (Essay on the Use of Leisure in Retirement) of AD 126, he says:

> Certain base scholars used to report evil of me to the emperor, but
> I decided not to worry about such affairs, or to learn their 'unique
> arts' (of civil service intrigue). Yet linked wheels may be made
> to turn of themselves, so that even an object of carved wood may
> be made to fly alone in the air. With drooping feathers I
> have returned to my own home; why should I not adjust my
> mechanisms and put them in working order (so that I may fly
> still higher than before)?

Here then he seems to mention his own mechanical interests, using them as an analogy for his own situation out of office.

The chief Western parallel to this is the 'flying dove' of Archytas of Tarentum, more or less of a Pythagorean who flourished in the neighbourhood of 380 BC, making him a contemporary of Mo Di and Gongshu Pan. Unfortunately, there is little reference to his model aircraft earlier than about AD 150, and so contemporary with Zhang Heng. One description says that it flew by means of some expanding vapour. Other accounts suggest that a weight and pulley were involved, and while the object could fly it could not rise again after falling. This might suggest that a launching mechanism was used, after which the model went forward in gliding flight assisted by whatever power source is implied by the reference to compressed air or steam. The invention seems much more in the Alexandrian manner than of the time of Archytas. Certainly the Alexandrians were concerned with pneumatic devices.

The kite and its origins

Let us now examine more closely the chief material basis for Chinese aeronautical stories, the kite of wood, bamboo and paper. Its use in Asia would seem to be exceedingly old, since anthropologists have found it in a wide distribution radiating south and east of China through the Indo-Chinese culture-area, as well as Indonesia, Melanesia and Polynesia. In some parts kite-flying was practised as a religious function connected with gods and mythical heroes. Often tabooed to women, the kite frequently carried, as in China, attachments such as strings or pipes to make musical or humming noises in the air. An important practical application was found for it in fishing to remove the hook and bait far from the sinister shadows of the boat and the fishermen. In China a game was played with kites; the cords were covered with crushed glass or porcelain glued on, the players seeking to get their kites windward of their rivals so that they could sever the cords.

The origin of the kite lies so far back in Asian history that theories about it – for instance, that it derived from shooting off arrows with cords attached, or from the bull-roarer which is whirled round on a string and was used as a ritual object in many primitive cultures – can be but speculative.

We may well be prepared to regard the devices of Mo Di and Gongshu Pan as the earliest references to kites, though some would not agree. However, scholars of the Song, notably Gao Cheng [Kao Chhêng] and Zhou Daguan [Chou Ta-Kuan] (thirteenth century AD) recorded a story that the Han general Han Xin [Han Hsin] (died 196 BC) flew a kite over the palace to measure the distance which his sappers would have to dig in order to make a tunnel through which his troops could enter. Later, in the Tang we learn that at the siege of Nanjing [Nanking] in the Taiqing [Thai-Chhing] reign period (AD 547 to 549), many kites were flown by the defenders to communicate news of their plight to army leaders at a distance. Again, in the thirteenth century at the famous siege of Kaifeng we hear of the besieged sending up paper kites with writing on them, then releasing them when over the Mongol lines so that the Jin [Chin] prisoners there received messages inciting them to revolt and escape. Thus we have the first instance of a 'leaflet raid'. The

Chinese certainly made continuous military use of kites, perhaps lending additional plausibility to the original association with Mo Di, whose disciples were interested in military techniques.

Kite-flying as a pastime also goes back a long way. There are pictures of it in the Dunhuang frescoes from AD 698 onward. Literary descriptions occur in the tenth century AD and frequently in the Song and Ming, though the practice of fitting Aeolian harps (i.e. those played by the wind) may have started in the Tang or before, for it is closely associated with the famous maker of kites in the tenth century, Li Ye [Li Yeh], and Song references to this practice are numerous.

A kite is supported on the wind by a combination of three forces: its weight, the resistance of the air, and the compensating tension of the string. According to the strength of the wind the kite moves in a circle of which the string is the radius, rising when the wind freshens and falling to a vertical position in which it can no longer remain suspended when the wind falls. The operator can then keep the kite airborne during the calm period by keeping the string taut and running. This keeps a sufficient air flow under its flat surfaces, just as an artificial airstream lifts the true powered aeroplane. In the eighteenth century some of the greatest European mathematicians devoted attention to the theory of kites, but already, during the centuries of the existence of the kite in China several interesting refinements had been introduced. For instance, a second cord permitting control of the angle of attack according to the wind strength was introduced, and surfaces were also cambered (Fig. 385), though we do not know whether the latter began earlier than the eighteenth century, at which time the first suggestions for curved aerofoils were made in Europe. However, a point which should not be overlooked is the historical relation between the kite, the aircraft and the sailing-carriage (p. 183); indeed, the kite might almost be considered as a detached sail of the sailing-carriage. At various times efforts were made, not without success,

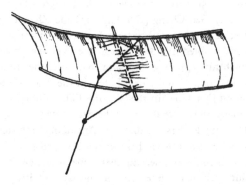

Fig. 385. Chinese cambered-wing kite.

to tow land-vehicles by means of kites, the most famous being that of a Mr G. Pocock on the road between Bristol and Marlborough in 1827.

The helicopter top; Ge Hong and George Cayley on the 'hard wind' and 'rotary wafts'

We now approach the most important part of this matter; the examination of the role which ancient and medieval Chinese aeronautical devices played in the basis for the vast modern development of aerodynamics and aviation. That the kite was unknown in Europe until the end of the sixteenth century AD, when it was brought back by early travellers, is well appreciated. The historian B. Laufer said that 'It makes its debut as a Chinese contrivance, and not as a heritage of classical antiquity", but this does not mean that it was unknown in the Islamic world; it was probably not new there in the ninth century AD when Abū 'Uthmān 'Amr Baḥr al-Jāḥiẓ described the flying of kites 'made of Chinese carton and paper' by boys. But in Europe the first description of kites occurs in Giambattista della Porta's *Magia Naturalis* (*Natural Magic*) of AD 1589. A few decades later they were employed in England for letting off fireworks in the air, while Athanasius Kircher, the Jesuit, whose relations with the China Jesuits were close, and who himself wrote on China, also refers to them in his *Ars Magna Lucis et Umbra* (*The Great Art of Light and Shadows*) of 1646 and states that in his time kites were made in Rome of such dimensions that they were capable of lifting a man.

All this is highly relevant to the developments in the nineteenth century. The study of kites did indeed in due course confirm experimentally their capacity to carry human astronauts aloft, but it was much more important in another way, because closely connected with the search for a suitable glider and aircraft wings. In 1804 Sir George Cayley constructed a successful model aircraft with flat kite wings and a tail rudder-elevator consisting of two flat kites intersecting at right angles. This was 'the first true aeroplane in history'. Flat surfaces attached to whirling arms were also used by him the same year for his fundamental physical experiments on air-resistance and other aerodynamic matters. Though the study of the wings of birds had long been proceeding in parallel, as early as 1799 Cayley had realised that the basic problem was 'to make a surface support a given weight by the application of power to the resistance of air'. An aircraft needed a supporting wing and a tail-unit to exercise control. Although Cayley understood that a cambered wing gave better lift, he did not feel compelled to build it into his full-size machines; in his model gliders he relied on the production of a curved surface by the airflow itself, acting as it did on his fabric wings which had spars only along the leading and trailing edges. But many of his drawings show the camber very clearly. Then, within fifty years, the conviction grew that one must imitate the cross-section of avian wings, and the majority of those who experimented with model flying-machines in the second half of the nineteenth century adopted curved aerofoil shapes of one kind or another in their wings. Long before, kite-makers in China had doubtless been led to this development by their preoccupation with the imitation of animal, especially bird, forms, and not from any aerodynamic considerations.

But still the paper bird of China had not exerted its full influence upon aircraft design, for in 1893 the Australian Lawrence Hargrave invented the box-kite for greater stability and lift, and it was this which inspired most of the biplane builders of the first decade of the present century. But before proceeding further, let us leap back in time some fifteen centuries, and pause to notice a very remarkable passage, on aerodynamics one might almost say, written by the great Daoist adept and alchemist Ge Hong [Ko Hung] about AD 320. This is what we find in the *Bao Pu Zi* [*Pao Phu Tzu*] (Book of the Preservation-of-Solidarity Master):

> Someone asked the Master about the principles of mounting to dangerous heights and travelling into the vast inane. The Master said . . . 'Some have made flying cars with wood from the inner part of jujube tree, using ox leather (straps) fastened to returning blades so as to set the machine in motion. Others have had the idea of making five snakes, six dragons and three oxen, to meet the "hard wind" and ride on it, not stopping until they have risen to a height of forty *li* [20,000 metres]. There the *qi* [*chhi*] is extremely hard, so much so that it can overcome (the strength of) human beings. As the Teacher says: "The kite (bird) flies higher and higher spirally, and then only needs to stretch its two wings, beating the air no more, in order to go forward by itself. This is because it starts gliding (literally, riding) on the 'hard wind'. Take dragons for example; when they first rise they go up using the clouds as steps, and after they have attained a height of forty *li* then they rush forward effortlessly (literally, automatically) (gliding)." '

For the beginning of the fourth century AD this is truly an astonishing passage, seemingly unequalled by any Greek parallel.

There can be no doubt that the first plan which Ge Hong proposes for flight is the helicopter top; 'returning (or revolving) blades' can hardly mean anything else, especially in close association with a belt or strap, for the device had radiating blades fixed at an angle to a rotating axis, which was set in rotation by pulling a cord (or strap) previously wound round the stem. One of its commonest names was the 'bamboo dragonfly', but in eighteenth-century Europe, this kind of toy was termed the 'Chinese top', though it seems to have been known in the West already in late medieval times. In 1784 it

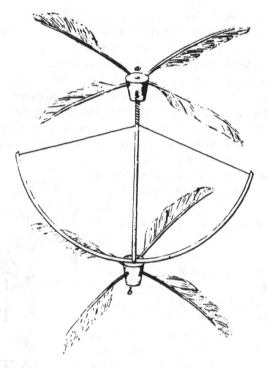

Fig. 386. The 'bamboo dragonfly' or Chinese helicopter top. A bow-drill spring rotates two feather airscrews (*a, b*) which carry the top high up into the air.

attracted the attention of Launoy and Bienvenu in France who made a bow-drill device to drive two contra-rotating propellers consisting of silk-covered frames (Fig. 386). In 1792 it stimulated Cayley to his first experiments on what he afterwards called 'rotary wafts' or 'elevating fliers'. He tells us so himself in 1809, and he too used a bow-drill spring to work two feather airscrews which kept the top mounting into the air. In 1835 Cayley remarked that while the original toy would rise no more than about 6 or 7.5 metres, his improved models would 'mount upward of 90 ft (27 metres) into the air'. This then was the direct ancestor of the helicopter rotor and the aircraft propeller.

There can be no doubt that the helicopter top was connected in its origin with the hot-air zoetrope (volume 2 of this abridgement, pp. 361 ff), and with the Mongol prayer-wheel operated by a chimney air current (p. 275). The use of a similar horizontal vane-wheel to work a roasting-spit, often found in the great kitchens of Europe, was apparently not known in East Asia. All were essentially rotors with vanes, moving in relation to currents of air parallel with their axes, the helicopter ad-aerially because of the motion imparted to it by the cord, the zoetrope and prayer-wheel ex-aerially because of the ascending current of hot air. However, since the aircraft propeller had to be vertically mounted, not only to bring about motor transportation (as the marine screw propels the ship), but to assure the airborne character of the flying-machine itself by driving the wings forward and so providing the necessary airstream lift, it would seem likely to spring from the European rather than the Chinese engineering tradition. For as we have already seen, Chinese technicians preferred horizontal mountings and Westerners vertical ones.

Certainly the role of the vertical windmill in the generation of the aeroplane propeller has considerable importance, yet the transition in position of the helicopter rotor to that of the aircraft propeller had already been made (at the ex-aerial level) in China. Liu Tong [Liu Thung], in his early seventeenth-century AD *Di Jing Jing Wu Lue* [*Ti Ching Ching Wu Lüeh*] (Descriptions of Things and Customs at the Imperial Capital) says that after kite-competitions were forbidden, many-coloured wind-wheels were made, which when set up facing the wind, or rapidly carried in the hand, whirled round showing their red and green colours confusingly. The arms of these wind-wheels, which were of course vertically mounted, were also used to do work as lugs depressing a lever and beating a drum.

In quite a different manner, moreover, Chinese technology had already prepared the way for those vertically mounted rotary roarers which would one day send the wings of aircraft tearing through the heavens. Earlier (p. 96 ff) we saw how advanced the medieval Chinese engineers were in their construction of rotary fans, notably the winnowing-fan used in agriculture, but also air-conditioning fans for palace halls. All these were vertically mounted just as propellers would one day be, and although they gave radial rather than axial flow, the rotary blowers of China preceded those of Europe by some fifteen centuries. Perhaps the most extraordinary prefiguration occurred when the Chinese toy-makers proceeded to fit Liu Tong's wind-wheels to children's kites.

Let us now return to Ge Hong. His words about the series of different kinds of creatures would be incomprehensible if we did not know well the perennial Chinese tradition of making kites in the shapes of animals. Dr Needham has no doubt that what he was referring to were man-lifting kites, and though we have as yet no evidence that Ge Hong or any of his contemporaries constructed such large devices, there would have been really nothing to prevent it. For people expert in kite-flying the possibility was obvious. And it happens that we do possess from a time not long after that of Ge Hong himself, a remarkable account of this very thing.

The setting was the reign of a cruel and tyrannical emperor, Gao Yang [Kao Yang] who ruled from AD 550 to 559. At a visit to the Tower of the Golden Phoenix, many prisoners condemned to death were harnessed with great bamboo mats as wings, and ordered to fly down to the ground from the top of the tower. Yet this was by no means the first time in Chinese history that trials had been made of wing-beating or ornithopter flight. As far back as the first century AD there was a well-authenticated attempt, while in AD 19, Wang Mang, the only Xin [Hsin] emperor, pressed by nomadic warriors from the north, put to practical test a man who claimed he could fly and so spy out the enemy. But in spite of 'rings and knots' connecting his wings, he flew only 'several hundred paces'. Ge Hong must certainly have known about this. One feels some difficulty in drawing any sharp distinction between the wing-beating 'tower-jumpers', as modern historians of aviation like to call them, and the eventually successful gliders, for some of the former (no doubt fortuitously) glided long enough before landing to survive. So although the first true glider flights were those of Cayley's passengers in 1852 and 1853, the idea has very ancient origins.

But there was something more interesting in the wicked emperor Gao Yang's proceedings of AD 559 than the imitation of birds. The *Zi Zhi Tong Jian* [*Tzu Chih Thung Chien*] (Comprehensive Mirror of History for Aid in Government) of the eleventh century AD, drawing upon other contemporary official sources, says:

> Gao Yang made Yuan Huangtou [Yuan Huang-Thou] and other prisoners take off from the Tower of the Phoenix attached to paper (kites in the form of) owls. Yuan Huangtou was the only one who succeeded in flying as far as the Purple Way, and there he came to earth.

Since the imperial road called the Purple Way was about 2.5km northwest of the city, Yuan Huangtou (a prince of Wei) managed in 'riding the hard wind' for a considerable distance. Moreover, the circumstances show that what was going on was not simply a cruel sport with prisoners, for the cables of the kites must have required manhandling on the ground with considerable skill, and with the intention of keeping the kites flying as long and as far as possible. Thus we have one circumstantial account of man-lifting kites within a couple of centuries of Ge Hong's time, and others are probably still buried in the texts. By the time Marco Polo was in China (*c.* AD 1285) man-lifting kites were in common use, according to his description, as a means of divination whereby sea-captains might know whether their intended voyages would be prosperous or not.

> And so we will tell you (he says) how when any ship must go on a voyage, they prove her business will go well or ill. The men of the ship will have a hurdle, that is a grating, of withies, and at each corner and side of this framework will be tied a cord, so that there be eight cords, and they will all be tied at the other end to a long rope. Next they will find some fool or drunkard and they will bind him on the hurdle, since no one in his right mind or with his wits about him would expose himself to that peril. And this is done when a strong wind prevails. Then the framework being set up opposite the wind, the wind lifts it and carries it up into the sky, while the men hold on by the long rope. And if while it is in the air the hurdle leans towards the way of the wind, they pull the rope to them a little so that it is set again upright, after which they let out some more rope and it rises higher. And if again it tips, once more they pull in the rope until the frame is upright and climbing, and then they yield the rope again, so that in this manner it would rise so high that it could not be seen, if only the rope were long enough.

If the hurdle makes for the sky, the augury is good, but if it has been unable to go up, the ship stays in port.

The wonders of modern aviation have thrown kites so much into the background that it is generally forgotten that they could ever supply sufficient lift to carry human beings into the air. Yet this development played its part in the history of aviation. A number of tentative trials took place from the time of Pocock in 1825 onwards, but full success was not attained until Hargrave invented the box kite in 1893. By 1906 it was possible for a man to remain for an hour at a height of over 700 metres suspended by a train of kites (Fig. 387). The significance of this was great. Only a few years before, Alexander Graham Bell had written: 'a properly constructed flying-machine should be capable of being flown as a kite, and conversely, a properly constructed kite should be capable of use as a flying-machine when driven by its own propellers'.

Lastly, what is to be said of Ge Hong's 'hard wind'? From the examples he gives of the gliding and soaring of birds, it is obviously nothing else than the property of 'air-lift', the bearing or rising of the inclined aerofoil subjected to the forces of an airstream, whether natural or artificial. (It will not be forgotten that we have met with the 'hard wind' of the Daoists before (volume 2 of this abridgement, p. 88) in Chinese astronomy in its role as a natural cause of planetary or stellar motion).

Fig. 387. A train of kites bearing aloft a military observer at the Rheims aeronautical meeting of 1909.

The birth of aerodynamics

Let us now try to place all this in correct perspective with regard to the growth of aeronautical science and practice. Leaving aside for the moment balloons and airships and jet-propulsion, we shall concentrate upon wings and air-screws. Man's attention (at least in the West) was attracted first by the beating of the bird wing, its gliding properties being neglected; hence Leonardo da Vinci's main interest was in flying-machines on the flapping or ornithopter principle. A decisive contribution came in AD 1681 when Alfonso Borelli showed that human muscles were anatomically and physiologically incapable of providing motive power for the unaided winged flight with the material then available, and there the matter rested. But the idea of beating wings was very tenacious. It was George Cayley at the beginning of the nineteenth century who broke completely with the old obsession and became a precursor of the Wrights rather than a successor to Leonardo. Cayley was the first to analyse the aerodynamic properties of the atmosphere by physical means, the first to lay down the scientific principles of heavier-than-air flight, the first to experiment with a captive flat surface at various angles to the flow of air, the first to make model and full-size gliders with rudders and elevators and to test them in free flight, the first to discuss streamlining and the 'centre of pressure' of a surface on an airstream, the first to realise that curved wings give a better lift than flat ones, and to recognise the low pressure area above them, the first to suggest multiple superposed wings, and the first to state that the lift of a flat surface varies in a specific way (the square of the relative air-speed multiplied by the density of the air). All this was done in a single decade, between 1799 and 1810. How great a pioneer Cayley was may be appreciated even further when we remember his studies on the airscrew, arising from the Chinese top; his anticipation of the internal combustion engine; and his practical and rational proposals for applying power to balloons. Finally, in 1903 came the first successful full-scale flying by the Wright brothers, using the internal combustion engine and an aircraft in all fundamental aspects identical with those of today. This incorporated both devices, the ad-aerial screw-rotor and the kite wing, which Ge Hong had spoken of sixteen centuries earlier. The kite was married to the windmill.

Such was the key contribution. Though the idea of it was implicit in Cayley's work, it remained 'in the air' during the first decades of the nineteenth century, crystallising in 1842-3 in W. S. Henson's famous design for 'an aerial steam carriage' which fitted airscrews to propel a fixed-wing aircraft. It would be interesting indeed if Chinese imagination participated in this crucial period, and we find that it did. In an 1832 reprint of a novel, the *Jing Hua Yuan* [*Chin Hua Yuan*] (Flowers in a Mirror) written by Li Ruzhen [Li Ju-Chen] between 1810 and 1820, and first published in 1828, there are 108 illustrations, two of which show imaginary flying-machines

Fig. 388. Drawing by Xie Yemei in 1832 showing three flying cars each with four screw-bladed rotors.

which combined propeller blades with kite surfaces resembling those of a biplane. The first depicts a vehicle open save for an awning travelling amongst clouds; it has four wheels as if for land travel, but between them on each side there is a screw-bladed rotor in a position similar to that of a wheel on a paddlewheel boat. The second is more interesting (Fig. 388), for it shows three flying vehicles each with four screw-bladed rotors taking the place of ordinary land wheels and, most curiously, between each of these propellers is a large gear-wheel which seems to connect them with a power source. This clearly shows that the artist had in mind mechanically driven ad-aerial (not ex-aerial) wheels. The vehicles themselves may echo children's kites fitted with toy wind-wheels which rotated as the kites flew. Whether or not his ideas derived from them, it seems that the illustrator, Xie Yemei [Hsieh Yeh-Mei] did really conceive of wheels acting in some way on the air and not merely rotated by it. So ended the classical contributions of Chinese culture to aeronautics.

The parachute in East and West

On account of its great simplicity, one would suppose that the parachute idea, like that of the sea-anchor, would be quite old in many civilisations. In Europe there seems to be nothing earlier than a description by Leonardo da Vinci about AD 1500 in the Codex Atlanticus, and apparently independently by Faustus Verantius about 1595. But historians doubt whether it was ever tried in practice before Jean-Pierre-François Blanchard and perhaps the Mont-golfier brothers used it for animals in 1778, and the human descents by Louis-Sebastien Lenormand and André-Jacques Garnerin some years later.

In China, however, there are much older references. In the *Shi Ji* [*Shih Chi*] (Historical Records), completed by 90 BC, Sima Qian [Ssuma Chhien] related a story about the legendary emperor Shun. His father, Gu Sou [Ku Sou] wanted to kill him, and finding him at the top of a granary tower, set fire to it, but Shun escaped safely by attaching a number of large conical straw hats together and jumping down. The eighth-century AD commentator Sima Zhen [Ssuma Chên] understood this clearly in the sense of the parachute principle, saying that the hats acted like the great wings of a bird to make his body light and bring him safely to the ground. A much later, but more circumstantial reference occurs in the *Ting Shi* [*Thing Shih*] (Lacquer Table History) written by Yo Ke [Yo Kho] in AD 1214. It concerns a report by the grandson of the great general Yo Fei describing what he saw in the foreign community of Arabs settled in Canton when his father was governor there. He tells of a golden cock on the top of a pagoda; this lacked one leg which had been stolen. The thief tried to sell the leg in the market and when questioned explained how he had hidden in the pagoda and sawn it off. Asked how he had then escaped, he replied 'I descended by holding on to two umbrellas without handles. After I jumped into the air the high wind kept them fully open, making them like wings for me, so I reached the ground without injury.'

This may be the tale of Shun and his father repeated later, with even the words preserved. Yet such indications must mean that the idea was current in China, while the observation of air-resistance to an outstretched fabric is so simple that it may have originated in many places independently. Indeed it would follow merely from the use of ship sails. If the parachute principle was not developed in China as it was in later Europe, this was because it was naturally ancillary to aviation itself, a typical piece of post-Renaissance technology.

Surprisingly, however, we have unusually concrete evidence that the invention was in fact at least once a transmission to the West. The Ambassador of Louis XIV in Siam, Simon de la Loubère, who was there in AD 1687 and 1688 described in his *Historical Relation* the exploits of Chinese and Siamese acrobats, saying:

> There dyed one, some Years since, who leap'd from the Hoop, supporting himself only by two Umbrella's, the hands of which were firmly fix'd to his Girdle; the Wind carried him accidentally sometimes to the Ground, sometimes on Trees or Houses, and sometimes into the River . . .

This passage was read in the following century by Louis-Sebastien Lenormand, who was stimulated to make practical trials from the tops of trees and buildings in AD 1783, which were quite successful. It was Lenormand who gave the parachute its present name and recommended it to Montgolfier, who fully appreciated its importance. This led to the descent of Garnerin from a balloon in 1797. There are not many cases in which so clear a line of transmission is detectable.

The balloon in East and West

The balloon is physically quite different from the parachute, for in one case the descent of a curved fabric surface is delayed by the drag of the air, while in the other, its ascent is facilitated by the presence of some medium lighter than air confined beneath it. The 'cloud captured in a bag' was a product of the use of the air pump in European pneumatic chemistry of the eighteenth century AD. In 1783 Pilâtre de Rozier and the Marquis d'Arlandes ascended in a Montgolfier hot-air balloon, and then in the following month in a hydrogen balloon an ascent was made by Jacques-Alexandre-César Charles and his mechanic, the elder of the Robert brothers who had built it. The simpler of these forms, using nothing but hot air, could have originated very much earlier than this, and in fact did.

Easter merry-makers in Europe in the seventeenth century AD had an entertaining trick of making empty eggshells rise in the air literally 'under their own steam'. Reported in many books, the procedure was simple enough. The contents of the egg were emptied through a small hole and the shell carefully dried, the right amount of dew (pure water) was introduced and the hole sealed with wax. Then in the hot sun the egg would move uneasily, grow light, and rise up into the air, floating for a moment before failing. How ancient this trick was in Europe we do not know, but in the second century BC the Chinese text *Huainan Wan Bi Shu* [*Huai-Nan Wan Pi Shu*] (The Ten Thousand Infallible Arts of the Prince Huainan) says 'Eggs can be made to fly in the air by the aid of burning tinder'. An ancient commentary incorporated in the text explains: 'Take an egg and remove the contents from the shell, then ignite a little mugwort tinder (inside the hole) so as to cause a strong air current. The egg will of itself rise in the air and float away." This method is more akin to that of the Montgolfier brothers than to that of the eggs raised by steam, since nothing but hot air was employed.

The discovery of this text puts a rather different complexion on the relations of China and Europe in the prehistory of flight, for Dr Needham is now inclined to think that the Han tradition was never lost. China was likely to be the home of hot-air balloons for several different reasons. Paper was available, as nowhere else in the world, from the Han period onwards and the development of the classical globe-shaped paper lanterns would have encouraged experimentation. When their upper openings were too small and the source of light and heat too strong, they must sometimes have shown a tendency to rise and float free of support. And indeed it is not difficult to find instances of the popular survival of hot-air balloons as an ancient sport in the Chinese culture-area. This century, in Lijiang [Lichiang] in Yunnan province in July, the critical month before the rainy season, the rice being already planted and the people having little to do, hot-air balloons are flown. Made of rough oiled paper pasted over a bamboo framework, with bunches of burning pine splinters underneath, they sail-up into the night air, some floating in the distance like red stars for several minutes before bursting into flames. There were similar pastimes in Cambodia. Medieval Chinese descriptions are still needed to fill the gap, but even so, this and the Han evidence together make a self-evident case for a perennial Chinese tradition, and indeed it is very unlikely that the tribal people and peasants of north-western Yunnan derive their proceedings from the France of Montgolfier.

It might even be urged that a practice originally Chinese was brought to the knowledge of Europeans at the time of the Mongol invasions. Much evidence has been collected from Eastern European chronicles that hot-air balloons shaped like dragons were used for signalling or as standards by the Mongol army at the Battle of Liegnitz in AD 1241. Certainly many of the fifteenth century German works on military technology, such as the manuscripts of Konrad Keyser's *Bellifortis* show drawings of horsemen holding what appear to be flying dragons in the air on the end of cords. He states that they contained oil-lamps as well as combustibles to give an effect of vomiting forth fire. In some respects these recall kites rather than hot-air balloons, but whatever the arrangements were, they seem to have continued into the next century. An account of the entry of Charles V into Munich in AD 1530, with an accompanying woodcut, attest the appearance of a similar flying or floating and fire breathing dragon.

If some of these creatures had an opening behind as well as at the front, they were perhaps precursors of the wind-sock. And here again there is an East Asian background, for some kites of Japan depict huge hollow paper fish with a large open mouth and a smaller opening at the rear. These were certainly used in 1886 for decoration and as sensitive wind-vanes or weathercocks, so no reverse influence from Western aviation technology could have been in play, and if the wind-sock was found in Japan it had almost certainly been in China earlier. Thus by tracing a tenuous thread, illuminated only very fitfully, we come back to a view often formerly held on less secure evidence, namely that China did play a considerable part in the prehistory of the balloon and the airship.

CONCLUSION

About the year 1911 an old gentleman taking a stroll in Beijing had his attention drawn to an aircraft flying overhead, but with perfect sangfroid remarked 'Ah, a man in a kite!' Chinese reactions to modern technology did not stop there, however, and quite a number of authors, though lacking that balanced judgment which an exhaustive acquaintance with sources both Eastern and Western alone could give, did not fail to maintain the emergence of occidental technology from oriental origins. About 1885, Wang Zhichun [Wang Chih-Chun] in his *Guo Chao Rou Yuan Ji* [*Kuo Chhao Jou Yuan Chi*] (Pacification of a Far Country) – an account of his ambassadorship to Russia – wrote:

> The useful arts and techniques originated from the earliest generations; thus geometry was invented by Ran Zi [Jan Tzu] (one of the disciples of Confucius), but later on the Chinese lost his books and Western people studied them, so that they became skilled in mathematics. So also the automatically striking clock was invented by a (Chinese) monk, but the method was lost in China. Western people studied it and developed refined (timekeeping) machines. As for the steam engine, it really originates from (the monk) Yixing [I-Hsing] of the Tang, who had a way of making bronze wheels turn automatically by the aid of rushing water – all that was added was the use of steam and the change of name. As for firearms, they originated in the fighting at Caishi [Tshai-shih] in the time of Yu Yunwen [Yü Yün-Wên] (of the Song); when he

defeated the enemy by the aid of certain firearms called pilipo [*phi-li-po.*] Thus the wonderful techniques of Western people are all based on the remains of ancient (Chinese) inventions. How could they be cleverer than the Chinese people?

Wang Zhichun's reaction was no doubt considered chauvinistic, and when such statements reached the ears of the 'gentlemen of the world' at that time, they were laughed out of court. But the progress of sober history has swung the pendulum the other way, and there now appears to be more solid basis for such protests. Whoever has read through the foregoing pages will allow that the balance shows a clear technological superiority of the Chinese side down to about AD 1400. The only important basic machine they did not have was the continuous screw, yet for this the great development of pedals and treadles (unfamiliar in Europe) was no small compensation.

Historians are coming to recognise this. In 1932 Lefebvre des Noëttes drew up a table of medieval acquisitions and stated clearly that many of them had come from the east of Asia. And in the 1950s the historian Lynn White could write that medieval technology consisted not simply of the equipment inherited from the Roman–Hellenistic world modified by the ingenuity of Western Europeans; it embodied also vitally important elements derived from the northern barbarians, the Byzantine and Near East, and from the Far East. At the same time, approaching the matter from the point of view of economic history, M. Greenberg in his *British Trade and the Opening of China, 1800–1842* pointed out that 'until the epoch of machine production, when technical supremacy enabled the West to fashion the whole world into a single economy, it was the East which was the more advanced in most of the industrial arts'.

Let the last words come from the realm of Islam, for the Arabs were very well qualified to be impartial judges of engineers both European and Chinese. In the ninth century AD Abū 'Uthmān 'Amr Baḥr al-Jāḥiz] wrote: 'Wisdom hath alighted on three things; the brain of the Franks, the hands of the Chinese, and the tongue of the Arabs.'

BIBLIOGRAPHY

Agricola, Georgius (George Bauer), *De Re Metallica*, Basel, 1556.

Alley, Rewi (tr.), *Tu Fu; Selected Poems*, Foreign Languages Press, Peking, 1962 (selection by Fêng Chih).

Alley, Rewi, 'Thangshan and the Eastern [Manchu Dynasty Imperial] Tombs [near Chi-hsien in Hopei]', *Eastern Horizon* (Hongkong), 1963, **2** (no. 12), 39.

Anderson, L. C., 'Kites in China', *New China Review*, 1921, **3**, 73.

Anderson, R. H., 'The Technical Ancestry of Grain-Milling Devices', *Mechanical Engineering* (New York), 1935, **57**, 611; *Agricultural History* (Washington, D.C.), 1938, **12**, 256.

Bathe, G., *Horizontal Windmills*, privately printed Philadelphia, 1948.

Beaton, C., *Chinese Album* (photographs), Batsford, London, 1945.

Beckmann, J., *A History of Inventions, Discoveries and Origins*, 1st German edn, 5 vols., 1786 to 1805; 4th edn, 2 vols. tr. by W. Johnston, Bohn, London, 1846; enlarged edn, 2 vols., Bell & Daldy, London, 1872.

Bedini, S., 'Johann Philipp Treffler, Clockmaker of Augsburg' *Bulletin of the Nat. Assoc. Watch and Clock Collectors* (U.S.A), 1956, **7**, 361, 415, 481.

Bell, Alexander Graham, 'The Tetrahedral Principle in Kite Structure', *National Geographic Magazine*, 1903, **14**, 219; 'Aerial Locomotion.' *National Geographic Magazine*, 1907, **18**, 1.

Berg, G., *Sledges and Wheeled Vehicles*, Stockholm, 1935. (Nordiska Museets Handlingar, no. 4.)

Bernal, J. D., *Science in History*, Watts, London, 1954. (Beard Lectures at Ruskin College, Oxford.)

Bernard-Maître, H., *Matteo Ricci's Scientific Contribution to China*, tr. by E. T. C. Werner, Vetch, Peiping, 1935. Orig. pub. as *L'Apport Scientifique du Père Matthieu Ricci à la Chine*, Hsienhsien, Tientsin, 1935 (rev. Chang Yü-Chê, *Thien Hsia Monthly* (Shanghai), 1936, **3**, 538).

von Bertele, H., 'Precision Time-Keeping in the pre-Huygens Era', *Horological Journal*, 1953, **95**, 794.

Birch, T., *History of the Royal Society of London*, Millar, London, 1756.

Biringuccio, Vanuccio, *Pirotechnia*, Venice, 1540, 1559. Eng. tr. C. S. Smith & M. T. Gnudi, Amer. Inst. Mining Engineers, New York, 1942.

Bolton, L., *Time Measurement*, Bell, London, 1924.

Boyer, M. N., 'Mediaeval Suspended Carriages.' *Speculum*, 1959, **34**, 359.

Boyer, M. N., 'Mediaeval Pivoted Axles.' *Technology and Culture*, 1960, **1**, 128.

van Braam Houckgeest, A. E., *An Authentic Account of the Embassy of the Dutch East-India Company to the Court of the Emperor of China in the years 1794 and 1795 (subsequent to that of the Earl of Macartney), containing a Description of Several Parts of the Chinese Empire unknown to Europeans; taken from the Journal of André Everard van Braam, Chief of the Direction of that Company, and Second in the Embassy.* Tr. L. E. Moreau de St Méry. 2 vols., map, but no index and no plates; Phillips, London, 1798. French ed. 2 vols., with map, index and several plates; Philadelphia, 1797. The two volumes of the English edition correspond to vol. 1 of the French edition only.

Branca, Giovanni, *Le Machine*, Rome, 1629.

Britten, F. J., *Old Clocks and Watches, and their Makers.* 6th edn, Spon, London, 1932; new edn edited G. H. Baillie, C. Chitton & C. A. Ilbert, 1954.

Broke-Smith, Brig. P. W. L., 'The History of Early British Military Aeronautics.' *Royal Engineers Journal*, 1952, **66**, 1, 105, 208.

Brooks, P. W., 'Aeronautics [in the late Nineteenth Century]', art. in *A History of Technology*, ed. C. Singer *et al.* vol. 5, p. 391, Oxford, 1958.

Burford, A., 'Heavy Transport in Classical Antiquity', *Eastern Horizon* (Hongkong), 1960 (2nd. ser.), **13**, 1.

Burstall, A., *A History of Mechanical Engineering*, Faber & Faber, London, 1963.

Burstall, A. F., Lansdale, W. E. & Elliott, P., 'A Working Model of the Mechanical Escapement in Su Sung's Astronomical Clock Tower', *Nature*, 1963, **199**, 1242.

Cardan, Jerome, *De Rerum Varietate*, Basle, 1557.

Cayley, Sir George, *Aeronautical and Miscellaneous Notebook, c. 1799–1826*, ed. J. E. Hodgson, Heffer, Cambridge, 1933. (Newcomen Society Extra Publications, no. 3.)

Cayley, Sir George, *On Aerial Navigation.* Collected (but abridged) papers from *Journal of Natural Philos., Chem. and the Arts*, 1809, **24**, 164 and 1810, **25**.

Cayley, Sir George, 'Retrospect of the Progress of Aerial Navigation, and Demonstration of the Principles by which it must be governed.' *Mechanics' Magazine*, 1843, **38**, 263.

Chadwick, N. K., 'The Kite; a Study in Polynesian Tradition.' *Journal of the Royal Anthropological Institute*, 1931, **61**, 457.

Chatley, H., 'Far Eastern Engineering', *Transactions of the Newcomen Society*, 1954, **29**, 151. With discussion by J. Needham, A. Stowers, A. W. Skempton, S. B. Hamilton *et al.*

Childe, V. Gordon, 'Rotary Querns on the Continent and in the Mediterranean Basin', *Antiquity*, 1943, **17**, 19.

Childe, V. Gordon, 'Rotary Motion [down to 1000BC]', art. in *History of Technology*, vol. 1, pp. 187–215, ed. C. Singer, E. J. Holmyard & A. R. Hall, Oxford, 1954.

Childe, V. Gordon, 'Wheeled Vehicles [in Early Times to the Fall of the Ancient Empires]', art. in *A History of Technology*, ed. C. Singer *et al.*, vol. 1, pp. 716–29, Oxford, 1954.

Cobbett, L., 'Mediterranean Windmills.' *Antiquity*, 1939, **13**, 458.

Crombie, A. C., *Augustine to Galileo; the History of Science,* *+400 to +1650,* Falcon, London, 1952.

Crone, E., Dijksterhuis, E. J. & Forbes, R. J. (ed.), *The Principal Works of Simon Stevin,* Amsterdam, 1955- .

Dickinson, H. W., 'The Steam-Engine to 1830', art. in *A History of Technology,* ed. C. Singer *et al.,* vol. 3, pp. 168–98, Oxford, 1958.

Diebold, John, *Automation; the Advent of the Automatic Factory,* New York, 1952.

Dijksterhuis, E. J., *Simon Stevin,* The Hague, 1943.

Drachmann, A. G., 'The Plane Astrolabe and the Anaphoric Clock.' *Centaurus,* 1954, **3,** 183.

Drachmann, A. G., 'Ancient Oil Mills and Presses', *Kgl. Danske Videnskabernes Selskab* (Archaeol-Kunsthist. Medd.), 1932, **1** (no. 1). Sep. publ. Levin & Munksgaard, Copenhagen, 1932.

Forbes, R. J., *Man the Maker; a History of Technology and Engineering,* Schuman, New York, 1950. (Crit. rev. H. W. Dickinson & B. Gille, *Archives Internationales d'Histoire des Sciences* (continuation of *Archeion*), 1951, **4,** 551).

Goodrich, L. Carrington, 'The Revolving Book-Case in China', *Harvard Journal of Asiatic Studies,* 1942, **7,** 130.

Greenberg, M., *British Trade and the Opening of China, 1800–1842,* Cambridge, 1951.

Haddon, A. C., 'The Evolution of the Cart', In Haddon, A. C., *The Study of Man,* Murray, London, 1908, p. 161.

Hall, A. R., 'More on Mediaeval Pivoted Axles', *Technology and Culture,* 1961, **2,** 17.

Hall, A. R., 'Military Technology [in the Mediterranean Civilisations and the Middle Ages]', art. in *A History of Technology,* ed. C. Singer *et al.,* vol. 2, p. 695, Oxford, 1956.

Harrison, H. S., 'Discovery, Invention and Diffusion [from Early Times to the Fall of the Ancient Empires]', art. in *A History of Technology,* ed. C. Singer *et al.,* vol. 1, p. 58, Oxford, 1954.

Hart, I. B., *The World of Leonardo da Vinci, Man of Science, Engineer, and Dreamer of Flight* (with a note on Leonardo's Helicopter Model, by C. H. Gibbs-Smith), McDonald, London, 1961.

Hiscox, G. D., *Mechanical Movements, Powers, Devices and Appliances used in Constructive and Operative Machinery and the Mechanical Arts . . .* (an illustrated glossary), Henley, New York, 1899.

Hodgson, J. E., *The History of Aeronautics in Great Britain from the Earliest Times to the latter half of the Nineteenth Century,* Oxford, 1924.

Hommel, R. P., *China at Work; an illustrated Record of the Primitive Industries of China's Masses, whose Life is Toil, and thus an Account of Chinese Civilisation,* Bucks County Historical Society, Doylestown, Pa., 1937; John Day, New York, 1937.

Hoover, H. C. & Hoover, L. H. (tr.), *Georgius Agricola 'De Re Metallica',* translated *from the 1st Latin edition of 1556, with biographical introduction, annotations and appendices upon the development of mining methods, metallurgical processes, geology, mineralogy and mining law from the earliest times to the 16th century,* 1st ed. Mining Magazine, London, 1912; 2nd ed. Dover, New York, 1950.

Hopkins, A. A., *The Lure of the Lock,* New York, 1928.

Hudson, G. F., *Europe and China; A Survey of their Relations from the Earliest Times to 1800,* Arnold, London, 1931 (rev. E. H. Minns, *Antiquity,* 1933, **7**, 104).

Jope, E. M., 'Vehicles and Harness [in the Mediterranean Civilisations and the Middle Ages]', art. in *A History of Technology,* ed C. Singer *et al.,* vol. 2, p. 537, Oxford, 1956.

Kyeser, Konrad, *Bellifortis* (the earliest of the + 15th-century illustrated handbooks of military engineering, begun + 1396, completed + 1410), MS. Göttingen Cod. Phil. 63 and others.

Lanchester, G., *The Yellow Emperor's South-Pointing Chariot* (with a note by A. C. Moule), China Society, London, 1947.

Laufer, B., 'The Prehistory of Aviation' *Field Museum of Natural History* (Chicago) *Publications,* Anthropological Series, 1928, **18**, no. 1 (Pub. no. 253).

Launoy, M. & Bienvenu, M., *Instruction sur la nouvelle Machine inventée par Messieurs Launoy et Bienvenu, avec lacquelle un Corps monte dans l'Atmosphère et est susceptible d'être dirigé,* Paris, 1784.

Lecomte, Louis, *Nouveaux Mémoires sur l'État présent de la Chine,* Anisson, Paris, 1696. (Eng. tr. *Memoirs and Observations Topographical, Physical, Mathematical, Mechanical, Natural, Civil and Ecclesiastical, made in a late journey through the Empire of China, and published in several letters, particularly upon the Chinese Pottery and Varnishing, the Silk and other Manufactures, the Pearl Fishing, the History of Plants and Animals, etc. translated from the Paris edition, etc.* 2nd ed. London, 1698. Germ. tr. Frankfurt, 1699–1700.)

Moritz, L. A., *Grain-Mills and Flour in Classical Antiquity,* Oxford, 1958.

Moule, A. C., 'The Chinese South-Pointing Carriage.' *T'oung Pao,* 1924, **23**, 83.

Mumford, Lewis, *Technics and Civilisation,* Routledge, London, 1934.

Needham, Joseph, 'The Translation of Old Chinese Scientific and Technical Texts', art. in *Aspects of Translation,* ed. A. H. Smith, p. 65, Secker & Warburg, London, 1958 (Studies in Communication, no. 2); and *Babel: Revue Internationale de la Traduction,* 1958, **4** (no. 1), 8.

Needham, Joseph, *Classical Chinese Contributions to Mechanical Engineering,* Univ. of Durham, Newcastle, 1961 (Earl Grey Lecture).

Needham, Joseph, 'The Prenatal History of the Steam-Engine' (Newcomen Centenary Lecture), *Transactions of the Newcomen Society,* 2nd ed., 1964.

Needham, Joseph, Wang Ling & Price, Derek J. de S., *Heavenly Clockwork; the Great Astronomical Clocks of Medieval China,* Cambridge, 1986.

Nieuhoff, J., *L'Ambassade [1655–1657] de la Compagnie Orientale des Provinces Unies vers l'Empereur de la Chine, ou Grand Cam de Tartarie, faite par les Sieurs Pierre de Goyer & Jacob de Keyser; Illustrée d'une tres-exacte Description des Villes, Bourgs, Villages, Ports de Mers, et autres Lieux plus considerables de la Chine; Enrichie d'un grand nombre de Tailles douces, le tout receuilli par Mr Jean Nieuhoff . . .* (title of Pt. II: *Description Generale de l'Empire de la Chine, ou il ist traité succinctement du Gouvernement, de la Religion, des Mœurs, des Sciences et Arts des Chinois, comme aussi des Animaux, des Poissons, des Arbres et Plantes, qui ornent leurs Campagnes et leurs Rivieres; y joint un court Recit des dernieres Guerres qu'ils ont eu contre les Tartares*), de Meurs, Leiden, 1665.

des Noëttes R. J. E. C. Lefebvre, *L'Attelage et le Cheval de Selle à travers les Âges;*

Contribution à l'Histoire de l'Esclavage, Picard, Paris, 1931, 2 vols. (1 vol. text, 1 vol. plates). (The definitive version of *La Force Animale à travers les Âges*. Berger-Levrault, Nancy, 1924.)

Partington, J. R., *A History of Greek Fire and Gunpowder*, Heffer, Cambridge, 1960.

Price, D. J. de S., 'Clockwork before the Clock', *Horological Journal*, 1955, **97**, 810; 1956, **98**, 31.

Price, D. J. de S., 'An Ancient Greek Computer' (the Anti-Kythera calendrical analogue computing machine). *Scientific American*, 1959, **200** (no. 6), 60.

Pritchard, J. L., *Sir George Cayley, the Inventor of the Aeroplane*, Parrish, London, 1961.

Reuleaux, F., *Kinematics of Machinery; Outlines of a Theory of Machines* (tr. A. B. W. Kennedy from *Theoretische Kinematik*, Wieweg, Braunschweig, 1875), London, 1876. French. tr. by A. Debize: *Cinématique; Principes fondamentaux d'une Théorie générale des Machines*, Savy, Paris, 1877.

Robertson, J. Drummond, *The Evolution of Clockwork, with a special section on the Clocks of Japan, and a Comprehensive Bibliography of Horology*, Cassell, London, 1931.

Rolt, L. T. C., *Thomas Newcomen; the Prehistory of the Steam Engine*, David & Charles, Dawlish, 1963; McDonald, London, 1963.

Sarton, George, *Introduction to the History of Science*, vol. 1, 1927; vol. 2, 1931 (2 parts); vol. 3, 1947 (2 parts), Williams & Wilkins, Baltimore (Carnegie Institution Pub. no. 376).

Sayce, R. U., *Primitive Arts and Crafts*, Cambridge, 1933.

Simpson, W., *The Buddhist Praying-Wheel; a Collection of Material bearing upon the Symbolism of the Wheel, and Circular Movements in custom and Religious Ritual*, Macmillian, London, 1896.

Singer, C., Holmyard, E. J., Hall, A. R. & Williams, T. I. (ed.), *A History of Technology*. 5 vols, Oxford, 1954–8.

Sowerby, A. de C., 'The Horse and other Beasts of Burden in China.' *China Journal of Science and Arts*, 1937, **26**, 282.

Sturt, G., *The Wheelwright's Shop*, Cambridge, 1942.

Tissandier, G., (*a*) 'Les Ballons en Chine', *La Nature*, 1884, **12**, (pt. 2), 287(*b*) 'Les Aérostats Captifs de l'Armée française' *La Nature*, 1885, **13** (pt. 1), 196(*c*) 'Les Aérostats de la Mission française en Chine', *La Nature*, 1888, **16** (pt. 1), 186.

Wailes, R., 'A Note on Windmills [in the Middle Ages]' art. in *A History of Technology*, ed. C. Singer *et al.*, vol. 2, p. 623, Oxford, 1956.

Wailes, R., 'Windmills [from the Renaissance to the Industrial Revolution]' art. in *A History of Technology*, ed. C. Singer *et al.*, vol. 3, p. 89, Oxford, 1957.

Wang, Chen-To, 'Investigations and Reproduction in Model Form of The South-Pointing Carriage and the Hodometer (*Li*-Measuring Drum Carriage), *Historical Journal National Peiping Academy*, 1937, **3**, 1.

Ward, F. A. B., *Time Measurement*, pt. 1, *Historical Review* (Handbook of the Collections at the Science Museum, South Kensington), HMSO, London, 1937.

Wei Yuan-Tai, 'Chinese Kites; their Infinite Variety', *China Reconstructs*, 1958, **7**, (no. 3), 17.

White, Lynn, 'Tibet, India and Malaya as Sources of Western Mediaeval Technology', *American Historical Riview*, 1960, **65**, 515.

White, Lynn, *Mediaeval Technology and Social Change*, Oxford, 1962.

Wiener, N., *Cybernetics; or Control and Communication in the the Animal and the Machine*, Wiley, New York, 1948.

Willis, Robert, *Principles of Mechanism*, Parker, London; Deighton, Cambridge, 1841; 2nd edn, Longmans Green, London, 1870.

Wolf, A. (with the co-operation of F. Dannemann & A. Armitage), *A History of Science, Technology and Philosophy in the 16th and 17th Centuries*, Allen & Unwin, 2nd edn, revised by D. McKie, London, 1950.

Woodbury, R. S., *History of the Lathe, to 1850; a Study in the Growth of a Technical Element of an Industrial Economy*, Soc. for the History of Technology, Cleveland, Ohio, 1961 (Soc. Hist. Technol. Monograph Ser. no. 1).

Zonca, Vittorio, *Novo Teatro di Machini e-Edificii*, Bertelli, Padua, 1607 and 1621.

TABLE OF CHINESE DYNASTIES

夏	HSIA [XIA] kingdom (legendary?)		*c.* −2000 to *c.* −1520
商	SHANG (YIN) kingdom		*c.* −1520 to *c.* −1030
周	CHOU [ZHOU] dynasty (Feudal Age)	Early Chou [Zhou] period	*c.* −1030 to −722
		Chhun Chhiu [Chun Qiu] period	−722 to −480
		Warring States (Chan Kuo [Zhan Guo]) period 戰國	−480 to −221
	First Unification 秦 CHHIN [QIN] dynasty		−221 to −207
漢	HAN dynasty	Chhien Han [Qian Han] (Earlier or Western)	−202 to +9
		Hsin [Xin] interregnum	+9 to +23
		Hou Han (Later or Eastern)	+25 to +220
	三國 SAN KUO [S GUO] (Three Kingdoms period)		+221 to +265
First	蜀 SHU (HAN	+221 to +264	
Partition	魏 WEI	+220 to +265	
	吳 WU	+222 to +280	
Second	晉 CHIN [JIN] dynasty: Western		+265 to +317
Unification	Eastern		+317 to +420
	劉宋 (Liu) SUNG [SONG] dynasty		+420 to +479
Second	Northern and Southern Dynasties (Nan Pei chhao [Nan Bei chao])		
Partition	齊 CHHI [QI] dynasty		+479 to +502
	梁 LIANG dynasty		+502 to +557
	陳 CHHEN [CHEN] dynasty		+557 to +589
	魏 Northern (Thopa [Touba]) WEI dynasty		+386 to +535
	Western (Thopa) WEI dynasty		+535 to +556
	Eastern (Thopa) WEI dynasty		+534 to +550
	北齊 Northern CHHI [QI] dynasty		+550 to +577
	北周 Northern CHOU [ZHOU] (Hsienpi [Xienbi]) dynasty		+557 to +581
Third	隋 SUI dynasty		+581 to +618
Unification	唐 THANG [TANG] dynasty		+618 to +906
Third	五代 WU TAI [WU DAI] (Five Dynasty period) (Later Liang,		+907 to +960
Partition	Later Thang [Tang] (Turkic), Later Chin [Jin] (Turkic), Later Han (Turkic) and Later Chou [Zhou])		
	遼 LIAO (Chhitan [Qidan] Tartar) dynasty		+907 to +1124
	West LIAO dynasty (Qarā-Khiṭāi)		+1124 to +1211
	西夏 Hsi Hsia [Xi Xia] (Tangut Tibetan) state		+986 to +1227
Fourth	宋 Northern SUNG [SONG] dynasty		+960 to +1126
Unification	宋 Southern SUNG [SONG] dynasty		+1127 to +1279
	CHIN [JIN] (Jurchen Tartar) dynasty		+1115 to +1234
	元 YUAN (Mongol) dynasty		+1260 to +1368 *c.*
	明 MING dynasty		+1368 to +1644
	清 CHHING [QING] (Manchu) dynasty		+1644 to +1911
	民國 Republic		+1912

N.B. When no modifying term in brackets is given, the dynasty was purely Chinese. During the Eastern Chin period there were no less than eighteen independent States (Hunnish, Tibetan, Hsienpi, Turkic, etc.) in the north. The term 'Liu chhao' [Liu chao] (Six Dynasties) is often used by historians of literature. It refers to the south and covers the period from the beginning of the third to the end of the sixth centuries AD, including (San Kuo) Wu, Chin, (Liu) Sung, Chhi, Liang and Chhen. The minus sign (−) indicates BC and the plus sign (+) is used for AD.

In this table modified Wade–Giles transliteration is given first and Pinyin transliteration follows in square brackets.

INDEX

Zu Chongzhi [Tsu Chhung-Chih] (fifth
  century AD), south-pointing
  carriages, 191, 198
Zuo Zhuan [*Tso Chuan*] (Master

Zuoqui's Tradition of the Spring
  and Summer Annals)
bellows, 90
hand-carts described in, 164–5

Printed in the United States
By Bookmasters